Pipeline Engineering - Design, Failure, and Management

Edited by Sayeed Rushd and Mohamed Anwar Ismail

Published in London, United Kingdom

Pipeline Engineering - Design, Failure, and Management
http://dx.doi.org/10.5772/intechopen.102290
Edited by Sayeed Rushd and Mohamed Anwar Ismail

Contributors
Alin Dinita, Mohamed Azzam, Sumit Kumar Singh, Denisa Ştet, Dan Doru Micu, Levente Czumbil, Mariano Alarcón, Manuel Seco-Nicolás, Juan-Pedro Luna-Abad, Alfonso P. Ramallo-González, Ravikiran Kadoli, Mukund A. Arun Patil, Milad Rezvani Rad, Andre McDonald, Minggang Tang, Satoru Hashimoto

First published in London, United Kingdom, 2023 by IntechOpen
IntechOpen is the global imprint of INTECHOPEN LIMITED, registered in England and Wales, registration number: 11086078, 5 Princes Gate Court, London, SW7 2QJ, United Kingdom

British Library Cataloguing-in-Publication Data
A catalogue record for this book is available from the British Library

Additional hard and PDF copies can be obtained from orders@intechopen.com

Pipeline Engineering - Design, Failure, and Management
Edited by Sayeed Rushd and Mohamed Anwar Ismail
p. cm.
Print ISBN 978-1-83769-993-3
Online ISBN 978-1-83768-000-9
eBook (PDF) ISBN 978-1-83768-001-6

For EU product safety concerns:
IN TECH d.o.o., Prolaz Marije Krucifikse Kozulić 3, 51000 Rijeka, Croatia,
info@intechopen.com or visit our website at intechopen.com.

Meet the editors

Dr. Sayeed Rushd is an Assistant Professor in the Department of Chemical Engineering at King Faisal University, Saudi Arabia. He obtained his Ph.D. in chemical engineering from the University of Alberta, Canada. In addition to teaching, Dr. Rushd is heavily involved in multiple research projects in the fields of pipeline engineering, dynamics of multiphase flows, rheology, water treatment, and desalination. He has authored or co-authored more than 35 technical papers. He is a reviewer of reputed journals, such as *Chemical Engineering Science, Engineering Applications of Computational Fluid Mechanics, Journal of Petroleum Science and Engineering*, and *Journal of Clean Production*. Dr. Rushd is a member of the Society of Petroleum Engineers.

Dr. Mohamed Anwar Ismail is a highly accomplished Assistant Professor of Mechanical Engineering at King Faisal University, Saudi Arabia, currently on leave from his position at Zagazig University in Egypt. He possesses a wealth of experience in both teaching and research, particularly in the fields of pipeline engineering, biomass combustion and pyrolysis, fire dynamics, renewable energy, and emission reduction from diesel engines. He holds a Ph.D. in mechanical engineering from the King Abdullah University of Science and Technology (KAUST). Dr. Anwar has authored or co-authored over 20 technical papers and is a respected member of the academic community, regularly reviewing for various reputed journals. He is a member of the Combustion Institute (CI).

Contents

Preface

The pipeline system is an integral part of our current civilization. Across the world, it ensures the economic transmission of essential fluids to different industries including oil and gas production, food processing, residential and industrial buildings, water supply, pulp and paper mills, pharmaceutical/medical manufacturing, and chemical manufacturing. Pipeline transportation plays as significant a role in running the world economy as do our blood vessels in ensuring life. The successful design, construction, and operation of a pipeline require a balance to be struck between production, cost, safety, and integrity. This book provides an overview of the current state of the art in the design, failure, and management of pipeline engineering with a specific focus on the most important evidence-based developments in this important area.

This book is divided into two main sections. The first section examines pipeline failure analysis and ways to prevent it, while the second looks at pipeline design and management. Chapter 1 is an analysis of various pipelines in the oil and gas industry. Chapter 2 explores new de-icing technology to prevent burst pipes. Chapter 3 evaluates electromagnetic interferences that affect metallic pipelines, while Chapter 4 reviews anomalies that impact the structural integrity of pipes, with a specific focus on local elastic-plastic deformations. The mechanical behavior of offshore flexible pipelines that operate under symmetrical and complicated loads is examined in Chapter 5, and methods to control the isothermal free vibration of functionally graded pipes using Terfenol-D are discussed in Chapter 6. Chapter 7 illustrates the use of a HEATT platform to simulate forced laminar flow in pipes subjected to asymmetric external conditions. Chapter 8 contains a detailed review of the stability of nanofluids used in pipelines, and Chapter 9 highlights the significance of government regulations in supporting pipeline network construction. Overall, the book provides a comprehensive examination of the technological details that aid pipeline engineers in their decision-making process.

In addition to thanking the authors, we would like to extend our gratitude to the management and facilitators at IntechOpen for the successful publication of this book. Dominik Samardzija and Iva Simcic deserve special thanks for their significant contributions. We would also like to thank King Faisal University for supporting the voluntary contributions of the editors.

Sayeed Rushd and Mohamed Anwar Ismail
Department of Chemical Engineering,
King Faisal University,
Hofuf, Al Ahsa, Saudi Arabia

Section 1

Pipeline Failure Analysis and Prevention

Chapter 1

Failure Analysis of Pipelines in the Oil and Gas Industry

Mohamed Mohamed Azzam

Abstract

The term "failure" can be defined as the inability of a part or assembly to perform its intended function. Despite the significant technological advances, failure incidents frequently occur, thus, causing human and financial consequences. The failure analysis is a crucial engineering tool. It aims to avoid similar cases in the future, thereby preventing accidents, reducing economic losses due to stopping plant production and keeping the environment safe. Furthermore, the failure analysis contributes to redesign, solve manufacturing drawbacks, save money and time, and in some cases, prevents fatality and saves lives. Conversely, failures can also improve engineering practices; indeed, through analyzing failures and implementing preventive measures, significant advances have been obtained in the quality of products and systems. Moreover, a beneficial outcome of failure analysis has been improved codes and specifications governing materials, for instance, API, ASTM, and ASME. In the current chapter, the failure analysis methodology will be discussed in detail with practical examples to know how to perform analysis for any failure cases, particularly in the oil and gas industry.

Keywords: failure analysis, pipeline, corrosion, fracture, oil & gas industry

1. Introduction

The term "failure" can be defined as the inability of a part or assembly to perform its intended function [1]. Based on the simple definition of failure, we can understand that the part of the component is considered failed if it cannot perform its function perfectly for any reason, for instance, a change in dimensions, corrosion, fracture, and so on. Sometimes unspecialized think the part to fail must be broken or fractured, but this is not the case. In other words, each fracture is considered a failure; however, not every failure is considered a fracture. A fracture separates parts into two or more species in response to the applied or residual stresses.

The pipeline can expose to thinning due to erosion-corrosion damage; however, it is still in service; thus, the pipeline can be considered to have failed, although it is still in service without leakage. The thinning mechanism is a failure since the pipeline has lost service life. In other words, the pipeline was designed to serve a specific period; however, the thinning damage has shortened its lifetime, which means the pipeline lost some of its lifetime. The high-pressure gas (HPG) pipeline has been subjected to internal corrosion, which led to a localized metal loss. Thus, the HPG pipeline

IntechOpen

has been converted to transfer oil or low-pressure gas due to the fitness for service, which revealed the remaining thickness cannot withstand the high pressure; thus, the pipeline has lost its function to transfer the high-pressure gas.

Material failure can be divided into four types: distortion or plastic deformation, fracture, corrosion, and wear [1]. It is worth noting that two or more physical failures can occur in the same failed part. The root causes of the failure can be divided into three levels; physical roots, human roots, and latent roots. The physical roots can be divided into four categories: design deficiencies, material defects, manufacturing or installation defects, and service life anomalies [2]. The human roots include inadequate inspection and improper equipment installed. The latent roots are the cultural or organizational rules that lead to the human cause; it is not direct roots. The inadequate inspector training is an example of the latent root, where some companies consider the training courses as additional costs that need to be reduced. These companies fail to recognize that this reduction is reflected in the ability of individuals, leading to catastrophic events due to incompetent persons.

The poor design can play a significant role in some failure cases; for example, the pipeline can be designed with a low spot that accumulates the water and causes corrosion. Also, poor design can create a crevice location which accelerates corrosion in the form of crevice corrosion due to the different concentrations of oxygen inside and outside the crevice. Additionally, material selection can be the root cause of some failure cases, using inappropriate material to serve in a harsh environment.

For instance, of inefficient material selection, using 304 austenitic stainless steel in chloride containing environment can lead to severe pitting corrosion or stress corrosion cracking [3]. Moreover, the manufacturing defects in most failure cases play a significant role in the failure. Thus, these defects act as the origin of catastrophic damage. For example, the lack of fusion of manufacturing defects can be a location for crack initiation or a stress concentration.

Furthermore, environmental change can cause premature or unexpected failure before its lifetime. For example, increasing the fluid velocity inside the pipeline accelerates the corrosion rate through erosion damage. Also, if the pipeline is designed according to a specific value (i.e., max limit) of hydrogen sulfide (H_2S) and carbon dioxide (CO_2), the increase of this limit would lead to anticipated failure mechanisms such as Sulphide stress corrosion cracking and CO_2 corrosion, especially in high-pressure gas (HPG) pipelines. In addition, the change in operating conditions may play a crucial role in the failure of pipelines. For example, increasing the operating pressure beyond the design pressure can lead to overloading damage or corrosion by increasing the partial pressure of the corrosive species such as Oxygen (O_2), hydrogen sulfide (H_2S), and carbon dioxide (CO_2) [4]. Also, the inadequate inspection may lead to failure; for example, performing the visual examination without nondestructive examination (NDE) may skip and ignore fine cracks or internal defects that the visual inspection disables to detect. Moreover, the absence of monitoring has a crucial role in some corrosion cases, where different monitoring methods are used to monitor the corrosion rate in the pipeline, such as the corrosion coupon, sand probe, and bio-probe. Without the monitoring method, assessing the operating condition, especially the fluid corrosivity, is not feasible.

Moreover, human error can play a critical role in failure cases through incompetent persons. For example, after conducting the hydro test for repaired tanks or vessels, and during the drainage of the used water, the responsible person can cause collapse due to rapid drain rate or due to the closing of the vent. Also, the non-drain of the water (i.e., Missing to drain) used in the hydro test can cause catastrophic failure, mainly if the pipeline is not used directly after the hydro test and is left for some time.

The hydro test water inside the pipeline with stagnation condition will be suitable for SRB colonies to grow and cause corrosion damage, called microbiological induced corrosion (MIC) [5]. Therefore, the water used in the hydro test must be flushed and entirely drained if the pipeline will be used directly after the hydro test. However, suppose the pipeline will not be used directly after the hydro test; in that case, the pipeline must be mothballed through injection of multifunction chemical which contains a mixture of oxygen scavenger, corrosion inhibitor, and biocide.

Indeed, failure analysis aims to determine the causes or factors that have led to an undesired loss of functionality; this considers the failure analysis's direct benefit. Also, failure analysis is an engineering tool for enhancing product quality and failure prevention; this considers the failure analysis's indirect benefit [2]. Furthermore, the failure analysis contributes to redesign, to solve the drawbacks of manufacture, saving money, saving time, and in some cases preventing fatality and saving lives [6].

The failure analysis must be performed based on a scientific base and a standard methodology to identify the damage mechanism and determine the significant root causes; otherwise, the analysis outcome will be unexpressed about the actual root causes. Also, the wrong root causes may accuse factors that are far from playing any role in the failure, thus, taking unsuitable recommendations which can accelerate the failure. The false root causes due to poor and inefficient failure analysis methodology can be likened to accusing an innocent person of murder; however, the real killer is free.

From my point of view, failure analysis in the oil and gas industry, specifically in the offshore environment, is considered a crucial case for the failure analysis society. Since these cases enrich the knowledge and information, thus, saving the marine environment against the consequences of failure and causing disasters such as pollution and marine life death in addition to economic loss. In April 2010, in one of the most significant pipelines containment failure accidents, that is, Deepwater Horizon, it was reported that approximately 3.19 million barrels of oil spilled into the ocean [7] and polluted at least 11,200 km^2 of seawater [8], which was catastrophic to both the economy and environment.

The crude oil at offshore platforms is transferred to the processing plants onshore through pipelines, which are considered one of the safest and most effective ways to transport oil and gas, security and reliability of the transmission pipeline [9]. There are three types of pipelines: gathering lines, transmission lines, and distribution lines. Gas or crude oil gathering lines exist between a well and a treatment plant or collection point [2]. The offshore pipelines are much more critical due to their operational condition, inspection, repair difficulties, and environmental issue [10]. These pipelines are manufactured from carbon steel. The consequences of pipeline rupture could lead to loss of life, injury, fire, explosion, environmental pollution, economic loss, decreasing capacity, and increasing maintenance difficulty [11].

2. Failure analysis methodology

The standard methodology for the failure analysis will be discussed in detail for each step. Practical cases will support the steps. The following steps are common for most failure cases [2]:

1. Collection of background data

2. Preliminary examination of the failed part

3. Chemical analysis (Sludge–Water-Liquid)

4. Nondestructive testing

5. Mechanical testing

6. Chemical analysis of the material

7. Selection, identification, preservation, and cleaning of specimens

8. Macroscopic examination and analysis

9. Microscopic examination and analysis (electron microscopy may be necessary)

10. Selection and preparation of metallographic sections

11. Examination and analysis of metallographic specimens

12. Analysis of fracture mechanics

13. Determination of failure mechanism

14. Testing under simulated service conditions (special tests)

15. Analysis of all the evidence, formulation of conclusions, and writing the report (including recommendations).

It is worth pointing out that the above methodology steps can be applied to most components in the oil and gas industry, such as storage tanks, pressure vessels, and pipelines. In other words, the above steps can be considered a generic methodology for most facilities in the oil and gas industry. However, in the further discussion for each step, most examples of the practical cases of failure incidents will be confined primarily to the pipelines, whether liquid or gas services, since the book's topic is mainly concerned with the pipelines. Therefore, it will be helpful to give examples of pipelines specifically.

2.1 Collection of background data

The failure analysis methodology's first step is collecting the background data. The failure investigation should include gaining an acquaintance with all pertinent details relating to the failure. In this step, the failure analyst acts as the detective who investigates a crime case by collecting all available data since the useless information, according to the operators of the failed part, is considered very important to the failure analyst and can contribute to solving the case. The failure analyst only who can judge the importance of the data.

Preparing a checklist containing all the essential required data and questions you need to ask is recommended to do not to forget anything. The list can include: the drawing or the as-built, history of the anomalies for the failed portion, repair history, history of the service environment (i.e., oil/water/gas/multiphase), pigging schedule, type of pig, whether BIDI or foam pig, water analysis reports, gas analysis reports, and scale analysis reports, in addition to the interview with the persons who are responsible for the operating of the failed portion.

It is worth noting that the failure analyst must be decent and non-offensive during the interview with the persons responsible for the failed part in the field and not accuse or blame anyone; otherwise, most of these persons vanish the important data fearing the blame or the accusation.

2.2 Preliminary examination

The preliminary examination can be divided into two stages; the first is performing a site visit to the incident location before retrieving the failed portion. The site visit to the incident location aims to figure out the actual situation of the whole location (i.e., the platform or the plant), not only the failed portion. In some cases, the preliminary site visit to the failure location of the failed part was the clue. Sometimes, the site visit to the failure location revealed simultaneous works implemented at the moment of the incident; consequently, the root causes are highly believed to be external causes due to these operational activities, not the failed part's environment. This indicates that the analyst must enlarge the investigation area around the failed part. Gradually narrow it, especially when the damage is external, such as a dent, gouge, and scratch.

The second stage in the preliminary examination is a visual examination of the failed part that depicts the actual condition after the incident without any change. The visual inspection aims to detect abnormal features such as damage morphology, plastic deformation, cracks, thinning, erosion, dents, dimensions change, and scratches. During the visual examination, a magnifying glass can be used [12] to enlarge some significant features of the failed part, such as the origin of the crack, damage morphology, and crack arrest location. Using a ruler or measure tap is recommended to determine the aspects of the failure, mainly in the cracking or rupture incident before movement or cutting the failed part, as shown in **Figure 1**, which

Figure 1.
Measuring the dimensions of the rupture of 24-inch water injection.

Figure 2.
Failed 2-inch bleeding valve due to drop of a mechanical tool.

shows the crack's measurement process dimensions for rupture of 24-inch water injection pipeline.

Figure 2 shows a failed 2-inch bleeding valve disconnected from the main line of high-pressure gas (i.e., 1200 psi), and the short nipple connection between the valve and the main line was found in a bent shape. On-site inspection indicated scaffolding had been installed around the failure due to concurrent work to replace an 8-inch water injection line directly above the failed valve. This means that the main reason for a defective valve to bend is that a mechanical tool fell and hit the valve, whereas the failed valve is a free end with no support. This failure case is an ideal example of the importance of on-site inspections at the fault location. Without on-site inspections, investigators will come up with unclear root causes and thus make incorrect recommendations.

The visual examination of the failed part must include determining the origin from which the crack or the corrosion started. The origin can be manifested as a bulge, such as the overloading failures, or stress concentration location, such as the sharp edges and weldment. For example, **Figure 3** shows the origin of the rupture, which occurred in an 18-inch gas pipeline, where the origin appeared in a bulge shape; the spout-like form (bulge) is indicative of an overloading incident. Eventually, during a visual inspection, failure analysts can use failure morphologies to brainstorm and imagine expected failure cause scenarios.

Figure 3.
Illustrates the origin (bulge) of fast-running ductile rupture of 18-inch oil pipeline.

2.3 Chemical analysis (sludge: water-liquid)

It is a crucial step to obtain a sample from the failed portion directly after the incident. The collected sample can be sludge or liquid. The sludge sample will be analyzed to determine the predominant components, especially in the corrosion failure cases. The scale analysis can provide us with informative data about the damage mechanism. Additionally, samples from debris and water are obtained to perform sulfate-reducing bacteria (SRB) testing to determine the count of sessile and planktonic bacteria in the fluid.

If the predominant compound in the analysis is iron carbonate, the expected damage mechanism is CO2 corrosion. If the iron sulfide (FeS) is the dominant compound, the predicted damage mechanism is microbiologically induced corrosion (MIC) due to the microorganism activity of the sulfate-reducing bacteria (SRB) [13]. If the predominant component is the sand, the expected damage mechanism is erosion damage due to the abrasive particles of the sand. If the main compound is iron oxide, the anticipated damage mechanism will be Oxygen corrosion.

The complete water analysis, iron content, H_2S dissolved, Oxygen dissolved, and CO_2 dissolved are significant parameters in most failure cases, indicating how the environment can be harsh. Chlorides have a detrimental effect on the passive layer, destroying the protection formed by the corrosion inhibitor. Oxygen is one of the corrosive species in any corrosion reactions, which accelerates the corrosion rate; therefore, it is recommended to control the Oxygen dissolved to levels of 10 to 50 ppb (part per billion). With increasing the partial pressure of CO_2, the dissolved concentration in the water increase, thus, lowering the pH and increasing the corrosion rate [14]. **Figure 4** illustrates the scale sample collection from a failed 24-inch oil pipeline.

2.4 Nondestructive testing

Nondestructive testing is a helpful and essential tool in most failure cases. Ultrasonic testing (UT) is often used to measure the remaining wall thickness to calculate the maximum pressure that can be applied. Also, UT is used to detect internal defects, whether base metal or weldment, like porosity, cracks, and laminations.

Figure 4.
Collection of scale sample from failed 24-inch oil pipeline.

Figure 5.
UT Technique (a) calibration of the UT device, and (b) Conducting of UT examination.

Additionally, UT can give the location and size of the defects [12] and measure the remaining wall thickness, as shown in **Figure 5**.

The dye penetrant test (PT) is often used to detect the surface cracks and clearly show the damage's extent, as shown in **Figure 6** shows a fine crack in the fillet weld of the 2-inch line of the High-pressure gas. The magnetic particles test (MT) also performs the same function as the PT, detecting surface cracks.

2.5 Mechanical testing

It is a crucial step among the steps of the analysis to confirm the desired mechanical properties of the failed part according to the specification to facilitate the subsequent steps, especially the stress analysis step. Thus, the tensile test determines the yield strength, ultimate tensile strength, and elongation. Additionally, hardness testing is the simplest of mechanical tests; it can be used to assist in evaluating heat treatment [6]. Furthermore, the impact test is used to measure the toughness of the failed parts

Figure 6.
Crack detected in the fillet weld of 2-Inch HPG line by dye penetrant test (PT).

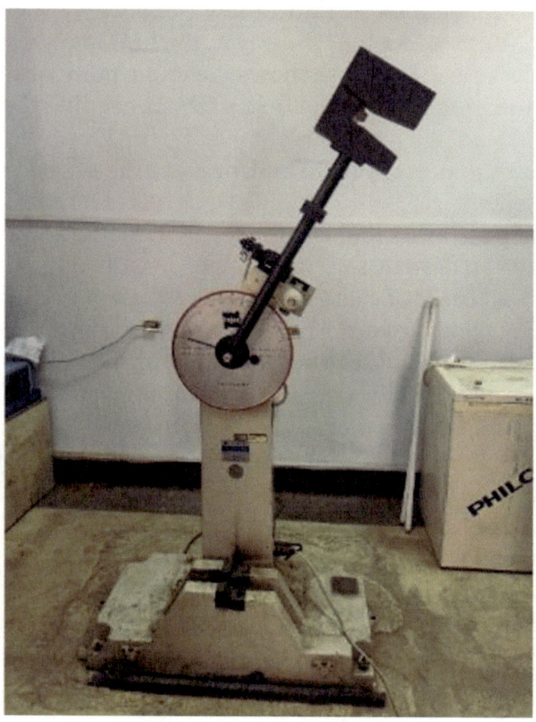

Figure 7.
Performing impact test for specimen cut from an 18-inch offshore pipeline.

in the overloading cases and fracture when the notch or point of stress concentration is experienced in the failure case [15]. The impact test is Mainly used when the failed part operates at low temperature or a mechanical tool hit or falls on it, as shown in **Figure 7**. It can be concluded that the mechanical tests are considered a verification that the material of the failed part was convenient to the applied stresses or the environment during the in-service period or not, according to the specification and the design criteria.

2.6 Chemical analysis of the material

In failure analysis cases, routine chemical composition analysis is highly recommended [6]. The chemical analysis of the material is used to determine the failed part's chemical composition, the same as the mechanical testing. The alloying elements in the cast alloys are rarely distributed uniformly. Thus, the chemical analysis is considered a verification tool by ensuring that the chemical composition does not deviate from the nominal composition according to specification. The chemical analysis is conducted in the laboratory using the optical emission spectrometer. The deviation from the nominal composition at a specific location of the failed part is called segregation [2]. **Table 1** illustrates the deviation of the chemical composition of 2205 duplex stainless steel pipe using X-Ray Fluorescence (i.e., XRF) device as shown in **Figure 8**, which caused a premature failure and gas release [16].

In some cases, it is difficult to perform the chemical analysis in the laboratory; therefore, positive material identification (PMI) is a prompt tool that can be used in the field to identify the material with the chemical compositions of the alloying element. The PMI is a portable device used to determine the chemical composition of the material without needing to transfer the sample to the laboratory. For example, **Figure 9** shows embrittlement in the flare of low-pressure gas, where the burner's designed material is 310 stainless steel; however, the PMI analysis revealed that the material is 384 stainless steel. This case shows the importance of the PMI and how it can facilitate determining the clue in the field without needing the laboratory.

2.7 Selection, sectioning, preservation and/or cleaning of specimens

In my view, the steps of selection, cleaning, and preservation are very critical and crucial since any fault can destroy the fracture surface, thus, making the failure analysis process difficult, and the root causes may not be present in the actual failure. The significant and valuable portion of the failed part is the origin, whether it is cracking or corrosion damage. The origin (i.e., the start point of the failure) is the clue of most failure cases since it would contain a defect, and the damage starts. **Figure 10** shows the locations to be cut for examination (i.e., mechanical testing, chemical testing, macro examination, and micro examination). The proficiency of the failure analyst is shown in the selection of what specific portions are to be studied. Since the whole failed part is not used in the analysis, small selected specimens, such as the origin, are expressed and valuable.

Furthermore, the sectioning step of the selected portion must be performed far enough from the failure's origin to prevent destroying it, which could lead to false

Element	C	Mn	Si	P	S	Ni	Cr	Cu	Mo	N
Pipe	0.013	0.8	0.61	0.038	0.001	4.69	22.5	0.06	3.49	0.06
ASTM A790 2205	0.03	2	1	0.03	0.02	4.5–6.5	22–23	.....	3–3.5	0.14–0.20
Flange	0.022	1.18	0.44	0.035	0.001	5	22.3	0.11	3.72	0.07
ASTM A182 F51	0.03	2	1	0.03	0.02	4.5–6.5	21–23	—	2.5–3.5	0.08–0.20

Table 1.
Chemical composition of 2205 Duplex stainless steel pipe.

DOI: http://dx.doi.org/10.5772/intechopen.108140

Figure 8.
Performing chemical analysis for the weldment using XRF.

Figure 9.
Embrittlement of flare due to unsuitable material.

conclusions [2]. During the sectioning process, it must be considered the sample size to be suitable for macro and micro examination and optical and scanning electron microscopes, respectively. It is recommended to use cold cutting techniques for

Figure 10.
Selection of location to be cut and studied.

Figure 11.
Cutting machine with cooling system.

sectioning the samples, like the wire cut and the abrasive blade cutting, which do not introduce any heat to the failed portion, thus, preventing any alteration to the actual condition, as shown in **Figure 11**.

Moreover, handling the selected portion of the failed part is a significant step. Therefore, the selected samples are stored in special boxes made from plastic to avoid friction between the specimens and the container. It is recommended to coat the pieces with grease or apply a removable coating of oil or plastic compound to prevent further interaction between the cut samples and the surrounding environment [12]. The surrounding environment causes corrosion and oxidation to the specimens and confuses whether the source of this corrosion is due to an in-service environment or not.

The cleaning process of the specimens is a crucial step since improper cleaning can destroy the fracture surface. The cleaning process aims to remove corrosion products, debris, and grease from the fracture surface. The cleaning process is an inevitable step

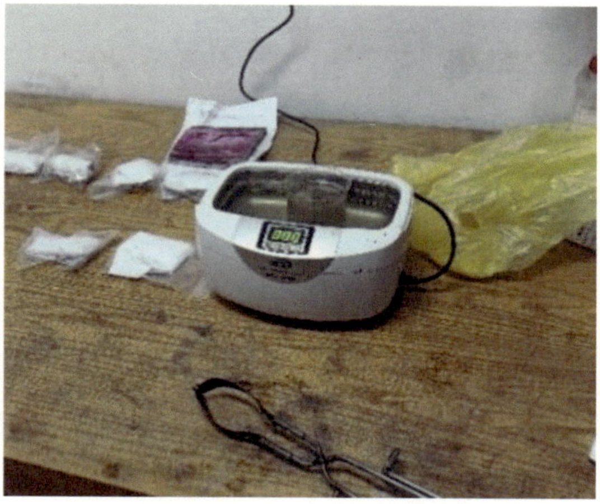

Figure 12.
Ultrasonic cleaning for specimen before examination by SEM.

in the microscopic examination, particularly when the scanning electron microscope is used. The cleaning program must be started with soft tools and gradually increase to aggressive cleaning tools based on the surface condition. Many cleaning methods can be used in the preparation of the samples. For instance, using the soft hair artist's brush as a preliminary step, and then using one or more methods from the following: using inorganic solvents, either by immersion or by jet, acetone or alcohol, cellulose acetate tape, replica, and use the ultrasonic cleaning bath. The ultrasonic bath is very useful in accelerating the cleaning process, as shown in **Figure 12**.

2.8 Macroscopic examination and analysis

The macro examination step is performed after the excellent cleaning of the sample in which the fracture surface becomes clear and free of corrosion products, dirt, grease, and residual species. The stereoscope is the most helpful tool in this step. The stereoscopic viewing has the same scope as visual inspection but is more detailed as typical stereoscopes allow 10X to 70X magnification [17]. The macroscopic examination does not require the fracture surface to be extremely smooth. The cleaning only is sufficient to perform a macroscopic analysis, unlike the microscopic examination (i.e., Optical microscope), which needs a polished surface to produce high contrast between the microstructural constituents.

The macroscopic examination can provide the failure analyst with beneficial information, for instance, the origin of the fatigue crack and secondary cracks, especially the fine cracks that could not be observed during the visual examination. Furthermore, the macroscopic shows a comprehensive view of the failure location compared to that of microscopic examination, thus, helping the failure analyst to imagine the scenario of what happened during the incident. **Figure 13** shows a cross-section macrograph of the failed pipe at the seam weld side. The joint configuration of the seam longitudinal weld is a double-V groove weld. A fusion welding process produces the seam weld. It is most likely a SAW process was used due to the considerable penetration depth in one pass (the filling pass). It is also evident that the crack propagated along the HAZ, HAZ/base metal boundary, or the base metal.

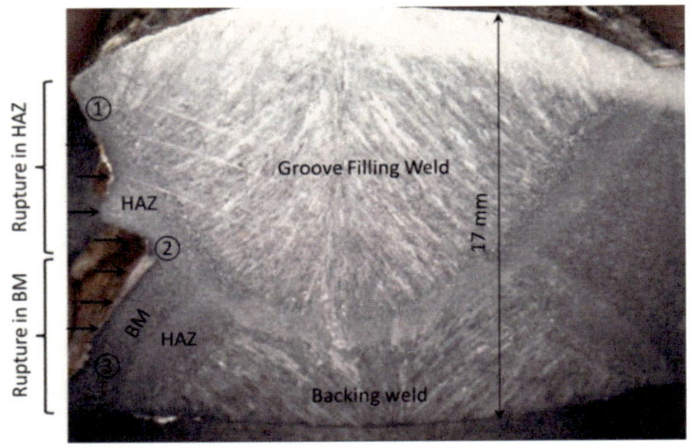

Figure 13.
Macrographs across the fracture surface at about 40 cm from the burst origin.

2.9 Microscopic examination and analysis

The microscopic examination is the clue in most failure cases, especially for metallurgical investigation, and is typically performed by the scanning electron microscope (SEM), as shown in **Figure 14**. The scanning electron microscope is an effective and helpful tool to know what happened during the incident; it looks like a recorded camera of the events of the failure. The scan electron microscope can provide a large magnification ranging from 5000 to 10,000X [6].

Some fracture modes can be identified according to the microscopic characteristics, where the dimpled morphology indicates ductile fracture due to overloading, and cleavage facets refer to brittle fracture. The striations are the most characteristic microscopic evidence of fatigue fracture. **Figure 15** shows cleavage facets at the origin

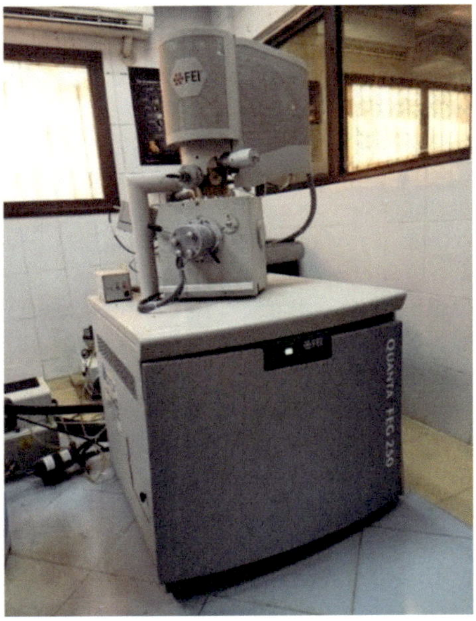

Figure 14.
Scanning electron microscope (SEM).

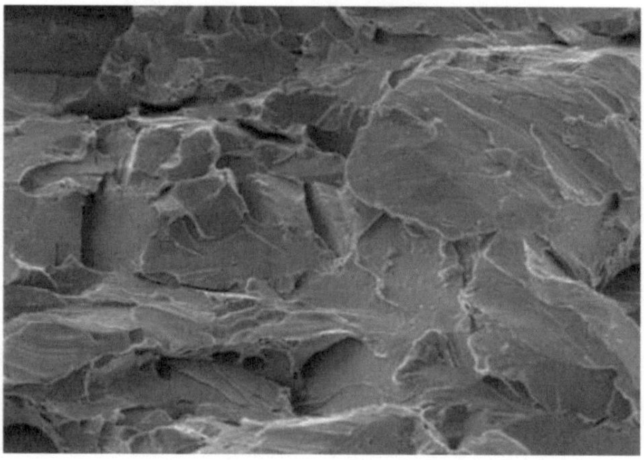

Figure 15.
Shows cleavage facets.

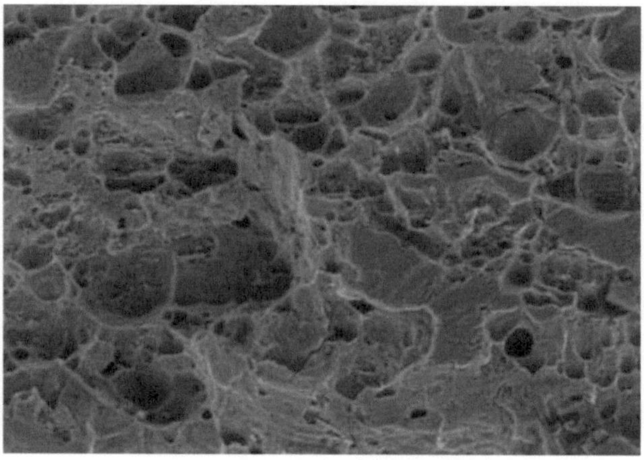

Figure 16.
Shows quasi cleavage.

of an 18-inch gas pipeline crack ruptured due to a welding defect. **Figure 16** illustrates the quasi cleavage of a fracture surface of an 18-inch oil pipeline, which ruptures in the form of rapid ductile fracture.

2.10 Selection and preparation of metallographic sections

The metallographic selection and preparation are crucial steps in metallurgical investigation cases. The selected region of the failed part must be chosen to present unique features of the failed part, which are selected for the characterization process. The selection of the samples which would be examined must be carried out carefully since these samples shall be near the edge of the fracture (i.e., around the origin). In other words, the significant sample which provides valuable information is the sample of the failure location. The sectioning of the metallographic specimens should be perpendicular to the fracture surface in edge view. Furthermore, selecting pieces far from the failure region (i.e., undamaged location) is helpful.

The selected specimens for the metallographic examination should be cut with cold manners. The cold cut prevents alteration of the specimens' microstructure, high precision cut, and deformation-free cutting for various workpiece sizes. Thus, the wire cut machine or the abrasive cut-off wheels are the standard practical methods for cutting the specimens examined in metallography. The samples are cut into small sizes to facilitate the handling and examination processes; these samples are ground using silicon carbide (SiC) foil and paper to produce a smooth surface before polishing.

2.11 Examination and analysis of metallographic specimens

Metallography is defined as the scientific discipline of examining and determining the constitution and the underlying structure of (or spatial relationships between) the constituents in metals, alloys, and materials (sometimes called materialography) [18]. The most common tool in the metallography examination is the optical or light microscope, with magnifications ranging from ~50 to 1000×, as shown in **Figure 17**. The optical microscope is used to identify the phases, constituents, and precipitations and determine the size and shape of the grains. The high contrast between microstructural constituents in light microscopy mainly depends on the quality of the specimen preparation process. The metallographic examination is also a verification of the heat treatment where the grains sizes are an indication of the heat treatment quality.

In **Figure 18**, it is evident that the HAZ at both sides of the joint contains coarse ferrite grains. In addition to the coarse ferrite grains, grain boundary austenite (GBA) has been observed along the grain boundaries of the coarse ferrite. It is believed that the excessive heat input of the welding process resulted in coarse-grained HAZ, which played a role in the degradation of HAZ zones.

2.12 Analysis of fracture mechanics (stress analysis)

It is sometimes quite apparent that excessive loading plays a detrimental role in the failure. Noticeable plastic deformation is observed at the failed pipeline due to

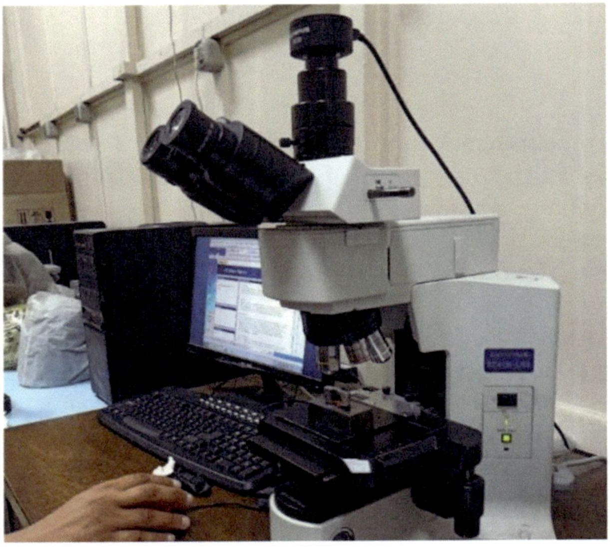

Figure 17.
The optical microscope for metallography.

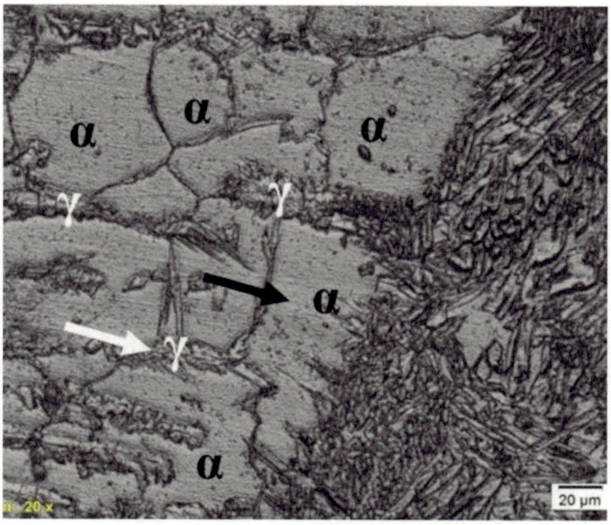

Figure 18.
Coarse grain HAZ of 2205 duplex stainless steel pipe.

the overloading, which causes a change in the pipeline configuration from circle to oval shape. Additionally, the stress analysis is performed based on the specifications, standards, and codes, for instance, ASME B31.8, API 579, and ASME 31G.

For the components with very complex shapes and high thermal gradients, a finite-element analysis (FEA) may be performed to estimate the most likely stress level in the failed part. These analyses can stand alone or can be used to help select critical locations for strain gauge attachment. Finite-element calculations can be time-consuming and expensive, but they are necessary for accurately assessing stress levels in areas

Figure 19.
Shows the topography of the ruptured wall and the wall bow-out.

of the complex geometry of some components. This analysis is almost essential for determining stresses caused by thermal gradients such as those found in welding [6].

Figure 19 shows an offshore pipeline ruptured due to overloading, where the outer surface of the ruptured wall is bowing out, indicating that the pipeline was highly inflated before it bursts. After failure, the surface is still bowing out, meaning that the deformation of the pipe wall was plastic deformation, and the level of stress encountered was very high.

2.13 Determination of failure mechanism

Determining the failure mechanism is almost the last step in the failure analysis methodology. The failure mechanism is determined based on the main finding of the visual examination, chemical analysis, mechanical testing, macro examination, and micro examination. It is believed that identifying the failure mechanism is easier and faster than determining the root causes of this failure damage. Since in most failure cases, the visual examination is preliminary and sufficient to figure out the failure damage.

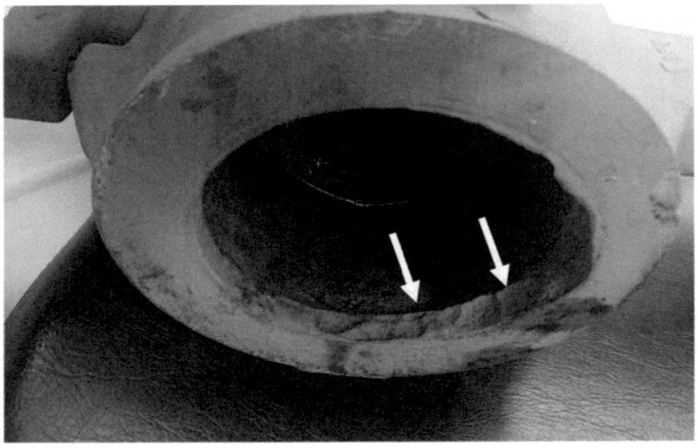

Figure 20.
Erosion at the 4-inch control valve of discharge line.

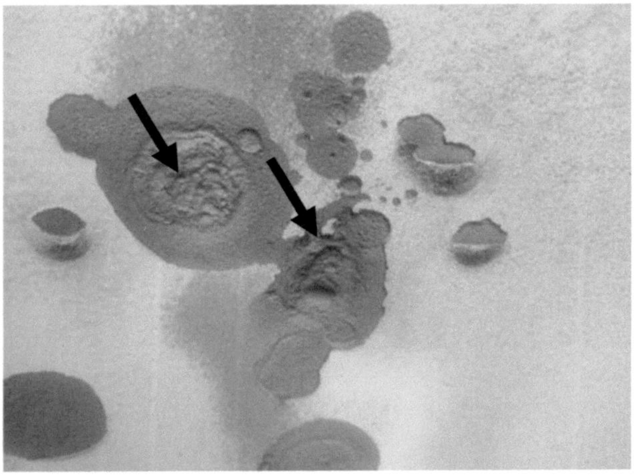

Figure 21.
Microbiological induced corrosion (MIC) at 30-inch discharge header.

Figure 20 shows a failed 4-inch control valve, and from the visual examination, it is clear that the expected damage is erosion-corrosion damage. **Figure 21** illustrates scattered pitting at the 30-inch oil discharge header. Based on the damage morphology and operating condition, it is believed that the microbiological induced corrosion (i.e., MIC) is the anticipated damage mechanism due to bacterial activity (i.e., SRB) and according to the cup-shaped damage.

Figure 22 shows a rupture of the 24-inch water injection pipe, where the observed bumps indicate an overload event; visual inspection shows severe thinning due to corrosion damage; therefore, with a working pressure of 1600 psi, the anticipated failure mechanism is Stress corrosion cracking. The corrosion cracking mechanism combines the stress and the corrosive environment [5]. Also, **Figure 23** shows a fracture in the 6-inch well flowline, and site visits to the platform show that the tubing is undergoing significant movement; therefore, the damage mechanism is believed to be fatigue damage due to cyclic loading.

Figure 22.
Stress corrosion cracking (SCC) at 24-inch water injection pipeline.

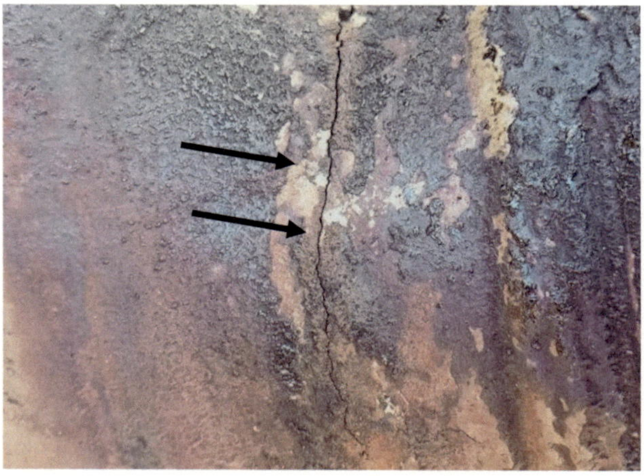

Figure 23.
Crack due to fatigue at 6-inch flow line of oil well.

2.14 Determination testing under simulated service conditions (special tests)

In some cases, simulation of operating conditions or special testing is strongly recommended to understand the effect of the environment on the same material of the failed component. It is important to note that most simulation tests are not feasible or practical because service conditions cannot be achieved, especially in the case of corrosion failure, which is not feasible in the laboratory. In addition, running simulations requires safety considerations, especially in the event of failures that occur under catastrophic conditions, such as overloading high-pressure gas lines. **Figure 24** shows pitting and crevice corrosion testing of an undamaged specimen cut from a defective 2205 duplex stainless steel weldment. This test is designed to determine the effect of the chloride (i.e., high salinity) on welded joints operating in harsh high salinity

Figure 24.
Pitting and crevice corrosion testing according to STM G-48.

Figure 25.
The specimen after conducting the pitting and crevice corrosion testing.

environments. The results show that preferential corrosion of the specimen occurs in the heat-affected zone and near the weld, as shown in **Figure 25**.

2.15 Determination analysis of all the evidence, formulation of conclusions, and writing the report

This step is the outcome of all stages in the failure analysis methodology. The main findings depict every step and are discussed and analyzed intellectually and scientifically to prove the damage mechanism and support the believed root causes. Also, all evidence is collected together during the investigation steps to complete the correct scenario about what has occurred.

The analyst's competence is also reflected in the preparation of the report, especially in the discussion and conclusion. Suppose the inspector has access to extensive laboratory facilities. In that case, best efforts should be made to analyze and discuss the results of mechanical testing, chemical analysis, fracture, and microscopy before formulating any preliminary conclusions [6]. Eventually, in investigations where the cause of failure is particularly elusive, searching for reports of similar cases may help identify possible root causes. Some references to other similar issues are suggested to support the discussion during the discussion.

3. Conclusion

The failure analysis methodology is considered the guide to perfect failure analysis in the oil and gas industry. Failure analysis in the oil and gas industry, specifically in the offshore environment, is a significant case for the failure analysis society. Since these cases enrich the knowledge and information, thus, saving the marine environment against the consequences of failure and causing disasters such as pollution and marine life death in addition to economic loss. It is helpful to publish the failure analysis cases and share them with others to enrich the knowledge and experience relevant to the society of failure analysis. The steps of the failure analysis methodology can be summarized as the following:

1. Collection of background data (i.e., Drawing, anomalies, repair history, environment, pigging activities, water analysis reports, gas analysis, and scale analysis)

2. Preliminary examination of the failed part (i.e., a site visit to the failure location and visual examination)

3. Chemical analysis (Sludge–Water-Liquid)

4. Nondestructive testing (i.e., UT, MT, PT, & RT)

5. Mechanical testing (i.e., Tensile test, impact test, & hardness test)

6. Chemical analysis of the material (Using the optical emission spectrometer &PMI)

7. Selection, identification, preservation, and cleaning of specimens (i.e., specimens' selection, wire cut, abrasive blade cutting, and ultrasonic cleaning)

8. Macroscopic examination and analysis (i.e., using of stereoscope)

9. Microscopic examination and analysis (i.e., Using of SEM)

10. Selection and preparation of metallographic sections (i.e., grinding, polishing, and etching of the samples)

11. Examination and analysis of metallographic specimens (i.e., optical microscope)

12. Analysis of fracture mechanics (i.e., stress analysis and finite element)

13. Determination of failure mechanism (i.e., identify the damage mechanism)

14. Testing under simulated service conditions (i.e., special tests)

15. Analysis of all the evidence, formulation of conclusions, and writing the report (including recommendations).

Author details

Mohamed Mohamed Azzam
Department of Metallurgical Engineering, Faculty of Engineering, Cairo University, Giza, Egypt

*Address all correspondence to: eng_azzam60@yahoo.com

References

[1] Wulpi DJ. Understanding How Components Fail. Second Edition. Materials Park, Ohio, USA: ASM International; 1999

[2] ASM International. ASM International. In: ASM Handbook. Failure Analysis and Prevention. Vol. 11. Materials Park, OH; 2005

[3] Davis JR. Corrosion of weldments. In: Chapter 3: Corrosion of Austenitic Stainless-Steel Weldments. Materials Park, Ohio, USA: ASM International; Dec 2006. pp. 43-75

[4] Bahadori A. Corrosion and Materials Selection. Chichester, West Sussex, United Kingdom: John Wiley & Sons, Ltd; 2014

[5] Winston Revie R. UHLIG's Corrosion Handbook. Hoboken, New Jersey: John Wiley & Sons, Inc; 2011

[6] Dennies DP. How to Organize and Run a Failure Investigation. ASM International; 2005

[7] United States of America v. BP Exploration & Production, Inc. et al. Findings of fact and conclusions of law: Phase two trial. In: RE: Oil Spill by the Oil Rig "Deepwater Horizon" in the Gulf of Mexico, on April 20, 2010. U.S. District Court for the Eastern District of Louisiana; 2015

[8] DWH-NRDA, Deepwater Horizon Natural Resource Damage Assessment (NRDA) – DRAFT. 2015 – section 4, injury to natural resources, DWH-NRDA. 2015. p. 685

[9] Han CJ, Zhang H, Zhang J. Failure pressure analysis of the pipe with inner corrosion defects by FEM.International Journal of Electrochemical Science. 2016;**11**:5046-5062

[10] Cheng YF. Stress Corrosion Cracking of Pipelines. Hoboken, New Jersey: John Wiley & Sons; 2013

[11] HSE. Guidelines for Pipeline Operators on Pipeline Anchor Hazards. Aberdeen: Health and Safety Executive; 2009

[12] McEvily AJ. Metal Failures Mechanisms, Analysis, Prevention. Hoboken, New Jersey: John Wiley & Sons, Inc.; 2013

[13] Javaherdashti R. Microbiologically Influenced Corrosion. London Limited: Springer-Verlag; 2008. pp. 49-100

[14] Palacios CA. Corrosion and Asset integrity Management for Upstream Installations in the Oil/Gas industry, CreateSpace Scotts Valley, California, 1st ed. 2016

[15] Callister WD Jr. Materials Science and Engineering-An Introduction. John Wiley & Sons, Inc.; 1985;**1985**

[16] Azzam M, Khalifa W. Investigation of duplex stainless steel flow line failure, engineering failure analysis (under review). 2022

[17] González-Velázquez JL. Fractography and Failure Analysis. Mexico: Springer International Publishing AG, part of Springer Nature; 2018

[18] ASM International. ASM Handbook. Metallography and Microstructures. Vol. 9. ASM International: Materials Park, OH; 2004

Chapter 2

Prevention of Pipes Bursting by Using a Novel Deicing Technology

Milad Rezvani Rad and Andre McDonald

Abstract

Freezing of water inside above-ground steel pipes is an unwelcome phenomenon that leads to internal pressurization, bulging, and bursting of pipes and can cause noticeable financial losses, environmental pollution due to the resulted leakages, and damage to the property, equipment, and workers in the field. Thus, a practical deicing/antiicing system that is highly efficient must be developed to minimize the detrimental impacts caused by freezing of liquids inside pipes. First, numerous tests were carried out in a relatively large cold room in which the actual working conditions of bare pipes exposed to cold weather were simulated to comprehend the freezing mechanism of the pressurized water. In the second phase of the project, the performance of the novel heating system was assessed by conducting deicing tests in the cold room. It was concluded that the freezing of the enclosed water was heavily dependent on the pressurization extent of water that itself was a function of pipe size and material properties. It was also found that the novel heating system that was produced by using thermal spraying means was able to eliminate the ice that was formed inside the pipe even under harsh conditions that may not be experienced in the field.

Keywords: deicing, functional thermal-sprayed coatings, joule heating, materials characterization, pipe bursting

1. Introduction

1.1 Steel pipe bursting

Ice formation inside steel pipes is a widespread incident in both residential and industrial sectors at locations where the ambient air temperature is below the freezing temperature of the water. When pipes are exposed to cold air for a prolonged duration, the ice that was formed initially at the inner surface of the pipe gradually accumulates and grows inward. In extreme cases, this causes a partial blockage inside a pipe or a full blockage at the thinner sections near connections and fittings. Once a portion of water has converted to ice, this causes an expansion in the volume that it occupies and when there is no further room in the pipe that can accommodate this volume growth, the pressure of the entrapped water starts building up. It is well-established that the pipe bursting occurs at a location where pressurized unfrozen water exists [1]. This incident is not only limited to the steel pipes, but it can also occur for plastic piping systems [2].

It is well-known that the negative consequences of bursting pipes are both costly and perilous. Furthermore, it can cause noticeable nonproductive downtime in industrial setups that results in loss of production. The financial losses because of this undesirable phenomenon are of great importance. According to the examination carried out by Insurance Information Institute [3], an average loss of $1.2 billion annually resulted from freezing issues only in the United States during the past 20 years and freezing of water inside pipes has always been one of the major problems.

In addition to the financial losses, the tremendous amount of energy that is stored in pipes is released after bursting of pipes. These explosive ruptures accompanied by a possible projection of chunks of steel from the damaged portions of pipes and pressure equipment can pose a threat to the well-being of the workers. Considering all the negative impacts of this undesirable phenomenon, it stands to reason that developing a novel and functional deicing system that can minimize or eliminate accumulation of ice inside pipes and pressure equipment is of great importance.

The common practice to minimize formation of ice is the installation of insulation around pipes. However, this cost-effective approach can only be practical in reducing the heat loss rate when pipes are exposed to cold ambient air for a short period of time [4]. That said, an efficient deicing system should accompany the thermal insulation to bring about satisfactory results when pipes are located in frigid environments for a prolonged period.

1.2 Usage of coating heating systems

It is well known that electric heat tracing is a practical method for deicing or temperature control of pipes that are used for water distribution, especially when waste heat and process steam are not obtainable [5]. In addition, heat tracers are employed where decent flow of high-viscosity fluids is needed. Another benefit of using electric heat tracing compared to steam-based tracing is its cleanliness [6]. Furthermore, they offer accurate temperature control and environmental and economic advantages in comparison to steam tracing [7].

One of the great features of electric heat tracing is the ability to control the amount of power that is supplied to the tracers. This can simply be done by input voltage adjustment. Thus, the heating or deicing operation can be adjusted at any given moment according to requirements in the field or the environmental conditions such as wind gust and air temperature. The system can become adaptable to climatic conditions and be operated automatically by using a controller [8]. The advantage of using controllers is twofold. Apart from the reduction in energy consumption, it can also increase the efficiency of the electrically resistive heating systems. Another feature of electric heat tracing is the wide range of power that can be supplied to tracers. In this regard, operating voltages that start from 24 VAC up to 750 VDC and power from 4 Watt per foot (W/ft) up to 30 W/ft at 10°C were reported [9].

An alternative to electric heat tracers is thermally-sprayed coating heaters that have been developed recently. These heaters are composed of several layers that have favorable mechanical and electrical features for the given tasks. The usage of these coating-based heaters has been investigated and tested for a wide variety of applications with different compositions, geometries, and power requirements [10–15].

1.3 Development and performance of coating-based resistive heating systems

It has been about a century since the thermal spraying concept was presented initially. This idea was optimized and refined over time, and it has become one of the most popular and cost-efficient means of additive manufacturing. This process employs a high-temperature flame or plasma to melt and accelerate tiny powder particles and deposit a relatively dense coating onto given surfaces. These coatings have been used successfully in different fields such as automotive industry and electricity production so much so that it has become a reliable and essential element in today's industry. Fabrication of electric resistive heaters by using thermal spraying processes has received an increasing attention recently.

The application of thermal spraying methods in production of meso-electronics was studied by Sampath, et al. [16]. It was concluded that different electrical elements including resistors, insulators, and conductors can be manufactured by using thermal spraying processes to deposit proper ceramic and metal feedstock materials that have favorable electrical properties. Nickel-chromium alloy was deposited onto alumina by using high-velocity oxygen fuel and plasma processes to develop resistors. The fabricated resistors had a sheet resistance in the range of 17–54 KΩ/sq. The advantages of this production procedure including high production rate, low cost of production, flexibility on thermal spray method, and ability to deposit millimeter-thick coatings made this method an interesting means of production of electrical components [16]. Similarly, high-quality conductors and dielectrics were developed by using both high-velocity oxygen fuel and cold spray processes [17].

1.3.1 Deposition of the electrically insulating layer

Thanks to its unique dielectric properties, alumina has been used widely as electrical insulating layer. Capability of thermal spraying processes in easy and quick deposition of alumina on large surfaces made this method appealing to experts in electronics industry. The volume resistivity of alumina coating fabricated by plasma spraying was reported to be very high in the order of 10^9–10^{10} Ω cm [18]. It was found that the electrical resistivity of alumina depends on several factors, namely humidity, microstructural characteristics, phase composition of alumina, and applied pressure [19–23].

1.3.2 Deposition of the heating element

The heating element was made by depositing a conductive metallic alloy that possesses relatively high electrical resistivity onto electrical insulating layer (alumina). Several researchers reported successful deposition and operation of several materials, namely molybdenum, nickel, nickel-20 wt.% chromium, nickel-5 wt.% aluminum, iron-13 wt.% chromium, and iron-chromium-aluminum that were manufactured by using different thermal spraying processes such as wire arc, VPS, APS, HVOF, and wire flame spray for usage as the heating element [10–15, 24]. As an example, three thermal spraying methods, namely vacuum plasma spray, air plasma spray, and high-velocity oxygen fuel, were used to develop heating elements to produce an easily controlled and uniform heat flux. The maximum heat fluxes that were generated from the developed samples before they fail were relatively high in the range of 10.6–17.2 MW/m^2 [10].

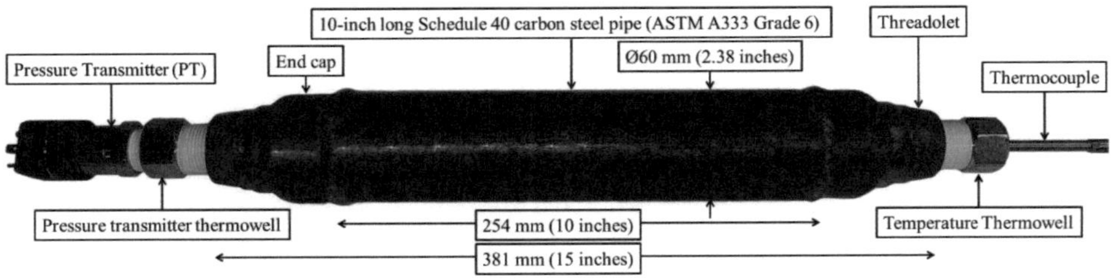

Figure 1.
Pipe assembly and the installed thermowells, pressure transmitter, and thermocouple used for freezing tests [25].

2. Bursting of steel pipes due to ice formation

Given solidification of water inside pipes and pressure equipment is a widespread phenomenon in both industrial and residential sectors, and it brings about negative consequences every year, placing emphasis on further study of the freezing mechanism of pressurized enclosed water seems reasonable. In this regard, numerous experiments were conducted on two well-known pipe materials, namely ASTM A106-B and ASTM A333-6, to study the freezing of pressurized enclosed water and examine the damage exerted on the integrity of pipes that led to their ultimate failure. Then, the fracture surfaces of the pipes were cut and removed from the pipes for further analysis by using a scanning electron microscope.

2.1 Experimental method

The pipe assemblies that were used for the freezing tests were composed of 254-mm long and 51-mm diameter carbon steel pipes. The samples were made from both ASTM A333-6 and ASTM A106-B materials so that the impact of pipe material on the freezing behavior of the enclosed water can be investigated. Schedule 40 pipes that are used extensively in industry were selected for these experiments. The geometry of the pipe assembly completed with fittings and sensors is shown in **Figure 1**.

The pressure and temperature data were collected regularly at the rate of one reading per second from the pressure transmitter and thermocouples. Two separate data acquisition systems (cDAQ-9171, National Instruments, Austin, TX, USA and SCXI-1600, National Instruments, Austin, TX, USA) were used for collection and storage of entrapped water pressure and temperature data, respectively. The NI MAX software package was used for collection of the measurements.

3. Development of multi-layered coating-based heater

A multi-layered thermally-sprayed heating system was fabricated to eliminate formation of ice inside the pipes and control the temperature of the enclosed liquid. Given the steel pipe was a conductive material, alumina layer as the insulating layer was deposited onto the carbon steel pipe to prevent flow of electrons and short-circuiting between the pipe and the conductive heating element. In addition to flame spraying process, cold spraying method was also used to deposit dense copper coatings where proper electrical connections were required. The effect of various spraying parameters on the quality and heating performance of the coating heater was investigated.

After the samples were fabricated and their heating functionality was tested, they were sectioned for microstructural evaluation and elemental composition analysis. Important features of the coating system, namely continuity, homogeneity, and microstructural defects, were analyzed by using a scanning electron microscope.

3.1 Experimental method

3.1.1 Feedstock powder

Several powder feedstock materials, namely aluminum oxide (Al_2O_3, AMDRY 6060, Oerlikon Metco, Westbury, NY, USA), nickel-50 wt.% chromium (50 Nickel-50 Chromium, 1260F, Praxair, Concord, NH, USA), and copper (SST-C5003, CenterLine, Ltd., Windsor, ON, Canada), were utilized to produce the coatings by using flame spraying and cold spraying methods. The micrographs that were taken in secondary mode by using a scanning electron microscope (Zeiss Sigma 300 VP-FE, Carl Zeiss Canada Ltd., Toronto, ON, Canada) from these powders at 500x magnification are shown in **Figure 2a–c**. The alumina powder had angular/blocky morphology due to their manufacturing process, which was fusing and crushing, and their size distribution was between 5 and 45 μm [26]. The size distribution of Ni-50Cr powder particles was from 22 to 53 μm and their spheroidal shape was because of the gas atomization process that was used for fabrication of this powder [27]. The minimum purity of the copper powder that was used in this research was 99.7% and its size distribution was between 5 and 45 μm [28]. The dendritic structure of this powder was obtained through electrolysis process.

3.1.2 Substrate preparation

The coating layers were sprayed onto the pipe sections so that the performance of the heating system as a deicing element can be assessed. The pipe sample that was used as the substrate in this study is shown in **Figure 3**.

In order to deposit the alumina layer onto pipe sections, the surface of the pipe was cleaned and then grit blasted with #24 alumina grit (Manus Abrasive Systems Inc., Edmonton, AB, Canada) to create the roughness required to cause the adhesion between the substrate and alumina coating. An air pressure of 586 kPa (85 psig) was used to accomplish the grit blasting process. Only the central section of the pipe assemblies was grit blasted and the ends were covered by using a heat-resistant masking tape (170-10S Red, Green Belting Industries, Mississauga, ON, Canada) to ensure that coating layers would not be deposited at the welded end caps.

3.1.3 Deposition of coating layers

Once the pipe assemblies were grit blasted, they were coated with a layer of pure alumina (Al_2O_3 99.5+ wt.%.). The powder particles were consistently fed to the oxy-acetylene torch (6P-II, Oerlikon Metco, Westbury, NY, USA). The spraying parameters for deposition of this layer were obtained through detailed trial and error procedure. Each time the microstructure of the obtained coatings was analyzed to ensure that the thickness of the coatings was uniform all over the substrate and that the coating layer was not damaged during the spraying process. Given the high melting point of alumina, relatively high flow rates for both acetylene and oxygen were selected to bring about a large high-temperature flame that is capable of melting the alumina powder particles. Furthermore, argon was selected as the carrier gas. The

Figure 2.
SEM images taken at 500 X magnification from powder materials, namely (a) Al_2O_3, (b) Ni-50Cr, and (c) Cu.

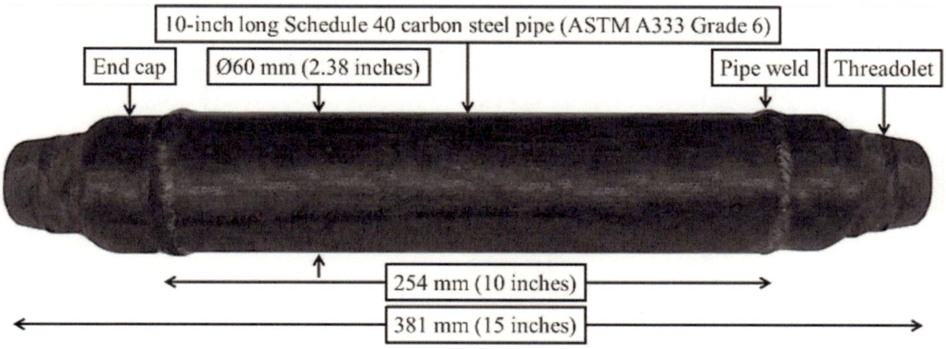

Figure 3.
The components of the pipe assembly that were used as the substrate [29].

flow rate of argon was set at 0.56 m^3/h (20 standard cubic feet per hour) to carry the powder particles to the flame spray torch that was installed on a programmable robot (HP-20, Motoman, Yaskawa Electric Corp., Waukegan, IL, USA). The robot was used to ensure the uniformity and reproducibility of the fabricated coatings.

The samples were put into a rotating chuck to create a uniform coating. As the torch was passed in a linear manner along the pipe, the pipe was rotated at the angular speed of 600 rpm. To avoid making helical patterns for the alumina coating, relatively low linear speeds of 10 mm/s for alumina layer, 24 mm/s for nickel-chromium layer, and relatively high angular speed of 600 rpm for the rotating chuck were selected. To minimize the inconsistencies, the powder feed rate was reduced, and the number of passes was increased to achieve the most uniform microstructure for the deposited coatings. A relative term (flow meter reading parameter) was used to adjust the flow of powder particles to the torch.

After deposition of alumina, nickel-chromium alloy with a composition of Ni 53 wt.%, Cr 46 wt.%, and Fe 1.0 wt.% was deposited onto the alumina layer by using flame spraying method. The structure of the pipe assembly complete with bi-layered coating can be seen in **Figure 4**. The alumina coating was intentionally sprayed over a longer section to ensure that conductive nickel-chromium coating would not be deposited directly onto the substrate as it can result in short-circuiting and malfunction of the deicing system.

In order to connect the developed coating system to the power source, copper coatings in the form of rings were fabricated at both sides of the pipe assembly. Copper was selected for this purpose because of its proper electrical properties, which

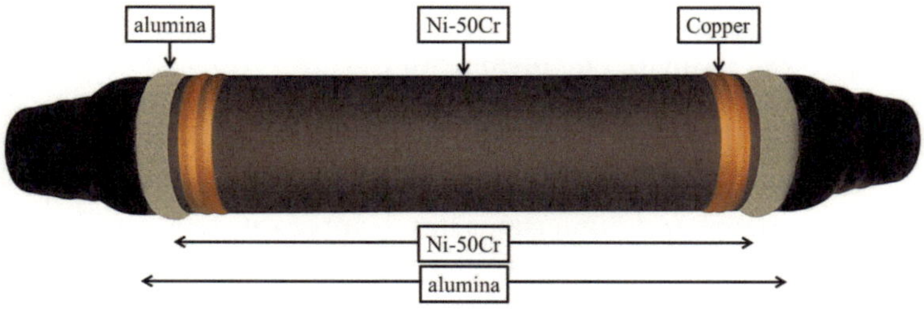

Figure 4.
The pipe assembly coated with flame-sprayed insulator and resistor layers and cold-sprayed connector [29].

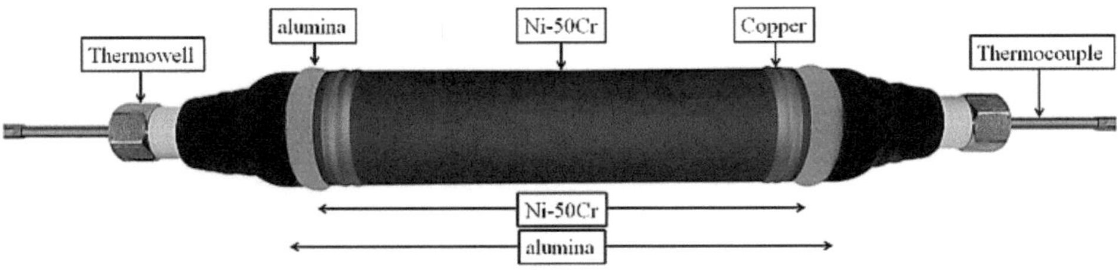

Figure 5.
The coated pipe assembly used for deicing tests completed with all fittings and thermocouples [30].

makes it an ideal option for electrical connections. The coating fabrication was accomplished by using the low-pressure cold spray system (SST series P, CenterLine, Ltd., Windsor, ON, Canada). The dendritic copper powder particles were accelerated inside the converging–diverging nozzle, and they formed a dense coating upon impact with the substrate. The working fluid that was used for this process to preheat and spray the copper particles was compressed air. The programmable robot was used for deposition of the copper rings as well. As the pipe was held in the chuck and was rotating, the cold spray nozzle passed the pipe assembly in the radial direction and generated the contact rings that are shown in **Figure 4**.

3.1.4 Installation of fittings and sensors

In order to read the temperature values of the enclosed water, two long thermocouples were inserted inside the thermowell at each end of the pipe assembly. The pipe assembly coated with the heating system is shown in **Figure 5**. In addition to the water/ice temperature, the temperature of the heating element was also measured by using a temperature sensor. The ambient air temperature inside the cold room was also measured by using a Type-T thermocouple.

3.1.5 Deicing test

The coated samples were put inside an 18.2 m^3 cold room freezer (Foster Refrigerator USA, Kinderhook, NY, USA) to simulate the harsh environmental conditions. The temperature of the cold room was set at -25°C with a $\pm$ 2°C bandwidth. To control the airflow over the pipe, the samples were placed horizontally inside a closed galvanized sheet metal duct with dimensions of 2 m $\times$ 0.66 m $\times$ 0.48 m. The impact of circulation of air inside the cold room was minimized for the free convection scenario by placing the lead of the duct over the pipe assembly. Furthermore, the forced convection was simulated by turning on the fan at the end of the duct. In this regard, a 0.25 kW (0.33 hp) direct-drive tube-axial fan (DDA-12-10033B, Leader Fan Industries, Toronto, ON, Canada) was used to simulate the harsh environmental conditions when pipes are exposed to wind gusts during cold winter days. Further details about the structure and dimensions of the apparatus that was designed and fabricated for this investigation can be found in the previous study [31]. **Figure 6** shows the location at which the coated pipe assemblies were placed during the freezing and deicing tests.

4. Results and discussion

4.1 Freezing behavior

Figure 7 shows the temperature and pressure traces that were obtained from the entrapped water inside Schedule 40 ASTM A333 Grade 6 pipe during the freezing test. The pipe assembly in this experiment was not fully filled with water and there was 5 vol.% of air inside the pipe. This graph demonstrates five distinct stages during the test, namely cooling of the water, the first solidification plateau, cooling of the mixture of water and ice, the second solidification plateau, and ice cooling. The water-cooling stage started when the fans inside the cold room started operating. This was followed by reduction in the temperature of ambient air inside the cold room. This stage was finished after a short super cooling period when plate-like solid crystals that are known as dendritic ice nucleated and grew [32]. Solidification of a portion of the enclosed water occurred in the next stage. The duration of this stage is directly impacted by the amount of air (gap) inside the pipe. It is well-known that the volume of water expands by 9% when it transforms into ice. Therefore, in case the vol.% of air inside the pipe is less than 9%, not all of the enclosed water can be converted to ice simply because there is not enough room for the volume expansion during the phase change. During this stage, the annular ice formed and grew inward [33]. Only a portion of water transformed into ice and once the pipe assembly was fully filled with the mix of water and ice, the pressure inside the pipe started building up. This noticeably reduced the rate of further water being transformed into ice because the pressure rise inside the pipe lowered the freezing temperature of water.

During the third stage, the pressure of the entrapped water sharply increased. This prevented the phase change of the rest of the entrapped water. A nonlinear relationship between melting temperature and pressure is obtained for hexagonal ice, which is given by Eq. (1) [34]. The purple curve in **Figure 7** shows the phase change pressure values for any given temperature according to this relation. As

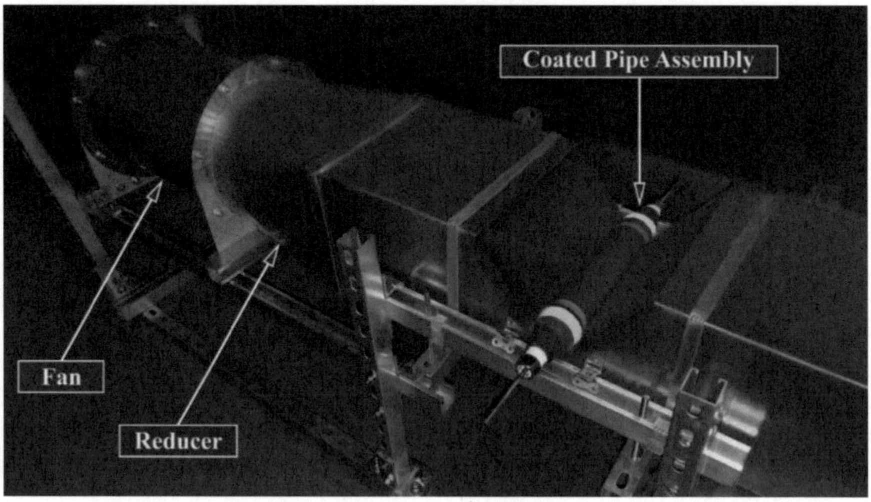

Figure 6.
The apparatus that was used for simulating the harsh environmental conditions [30].

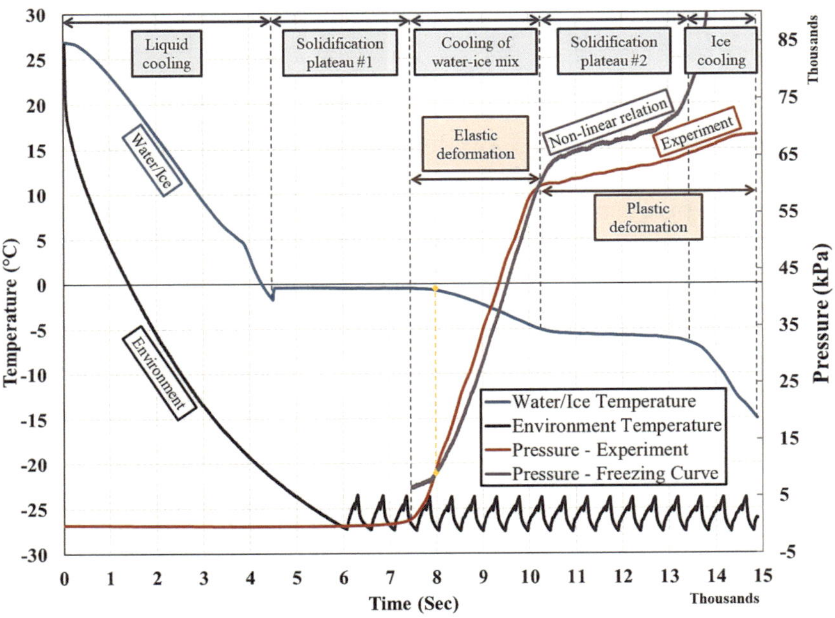

Figure 7.
Temperature and pressure traces versus time for water/ice inside the pipe assembly.

long as the pressure of the entrapped water was greater than the value obtained from this equation, water was not able to transform into ice even at subzero temperatures. This situation was discontinued when the pressure of the water was so high that it caused plastic deformation of the pipe. Afterward, given the volume of the pipe assembly was expanded during this inelastic deformation period and the pressure and the slope of pressure trace during this stage were much lower than that of the pressure trace during the elastic deformation stage, further water was able to undergo phase change process. Therefore, the second plateau that can be seen in fourth stage was indicative of the transformation of the remaining water inside the pipe into ice. Once the pipe was fully filled with solid ice at the end of the fourth stage, further exposure to the cold ambient air only reduced the temperature of the solid ice as shown in the fifth stage.

$$P_f = 6.11657 \times 10^{-4} - 414.5 \times \left[\left(\frac{T_f}{273.16} \right)^{8.38} - 1 \right] \tag{1}$$

It is well-established that the plastic deformation of the pipe occurs when the hoop stress in the pipe reaches the yield strength of the pipe material. The hoop stress itself depends on the pipe diameter. Therefore, it can be concluded that the freezing point depression and the freezing behavior of the water that depends on the plastic deformation of the pipe, itself is a function of the pipe size, wall thickness, and material. The hoop stress values can be obtained by applying the relations developed by thick-walled pressure vessel theory. It is worth mentioning that the yield strength of the pipe is not a constant value, and it changes from one freezing–thawing cycle to another cycle due to the work hardening of the pipe material that is caused by movement of dislocations.

The comparison of the pressure values obtained from the freezing experiment with the results obtained from Eq. (1) is only valid when still water is entrapped in the pipe. That being said, the increasing deviation between the pressure experimental measurements and theoretical estimations is due to the fact that only ice exists in the pipe by the end of the fourth stage, and therefore, that comparison is not valid anymore.

4.2 Plastic deformation and work hardening

A PI tape was used to measure the inelastic deformation of the pipes during the freezing experiment. In order to calculate the transverse strain of the pipe, the outer diameter of the pipes was measured before and after freezing experiments. Obviously, the maximum deformation took place in the middle of the pipe assembly where it was farthest from the circumferential welds and the reinforcing impact of the welds was at its minimum level. It was concluded that the maximum strain values that A333-6 and A106-B steel pipes could accommodate before failure were 11.3% and 13.3%, respectively. Although the minimum transverse elongation before rupture has been reported to be 16.5% for ASTM A106 Grade B steel [35], the pipe did not meet the expected value under freezing experiments. However, the obtained value for A333-6 steel was very close to the reported minimum transverse elongation, which is 11.4% [36].

The maximum pressure values inside the pipe were greatly impacted by the number of times the pipe underwent plastic deformation during the freezing and thawing cycles. The increase in the peak pressure value from one test to another test, which was because of the work hardening of A333-6 and A106-B Schedule 40 pipes, can be clearly observed in **Figure 8a** and **b**.

4.3 Pipe rupture

Figure 9 shows the failure pattern of the pipes after several freezing and bulging experiments, which is a common pattern for pipe bursting cases. The axial crack was formed in both pipes due to the developed excessive hoop stresses during the tests. It is well-established that the hoop stress is the maximum principal stress when pipes are pressurized. This pressurization was caused by gradual formation of ice accompanied by volume expansion and low compressibility of unfrozen water [1]. This pattern and location of the failure were expected as it is known that pipes fail where water freezes last. In this case, the remaining unfrozen water was entrapped in the middle of the pipe where final rupture occurred.

It can be seen in **Figure 9** that the extent of the deformation of the pipes was different for these two pipes. At the failure spot, the generated gap was wider and larger for the A106-B pipe compared to the A333-6 pipe. This was because of the amount of energy that was absorbed by the pipe material during the formation and propagation of the crack upon rupture. It is believed that the toughness of these pipe materials at low temperature has an important role to play. It is well known that the A333-6 pipe is the preferred option for installation at sites that are exposed to low-temperature environments because of its acceptable low-temperature toughness. Therefore, the higher toughness of A333-6 pipe compared to A106-B pipe results in greater resistance against brittle fractured caused by internal pressurization and overload. In contrast, the degradation of absorption energy of A106-B pipe material at lower temperature was found by conducting Charpy impact test [37].

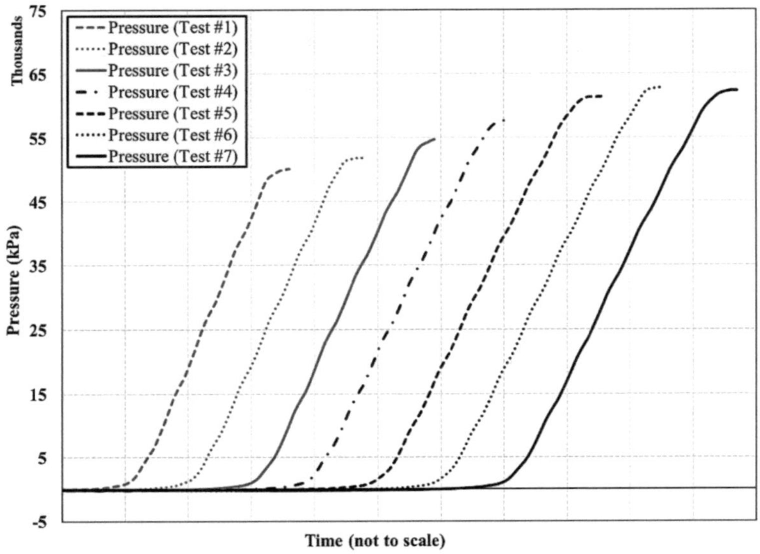

(a)

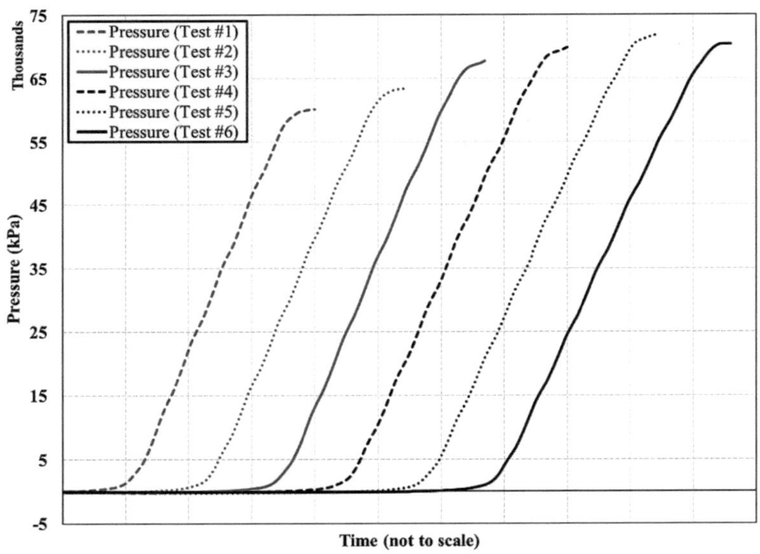

(b)

Figure 8.
The impact of work hardening of the pipe materials on the peak pressure during the freezing test for (a) A106-B and (b) A333-6 pipes.

In order to observe the fracture surfaces, the pipes were cut after the failure. To study the fracture patterns at different locations of the pipes, eight different rings were cut at 1 cm intervals from both pipes as shown in **Figure 9**.

4.4 Failure analysis

High-quality side images were taken from the rings shown in **Figure 9** to observe the macroscopic fracture features. These features, along with the geometry of the fractured surfaces are often used to determine the mechanism of the fracture and also

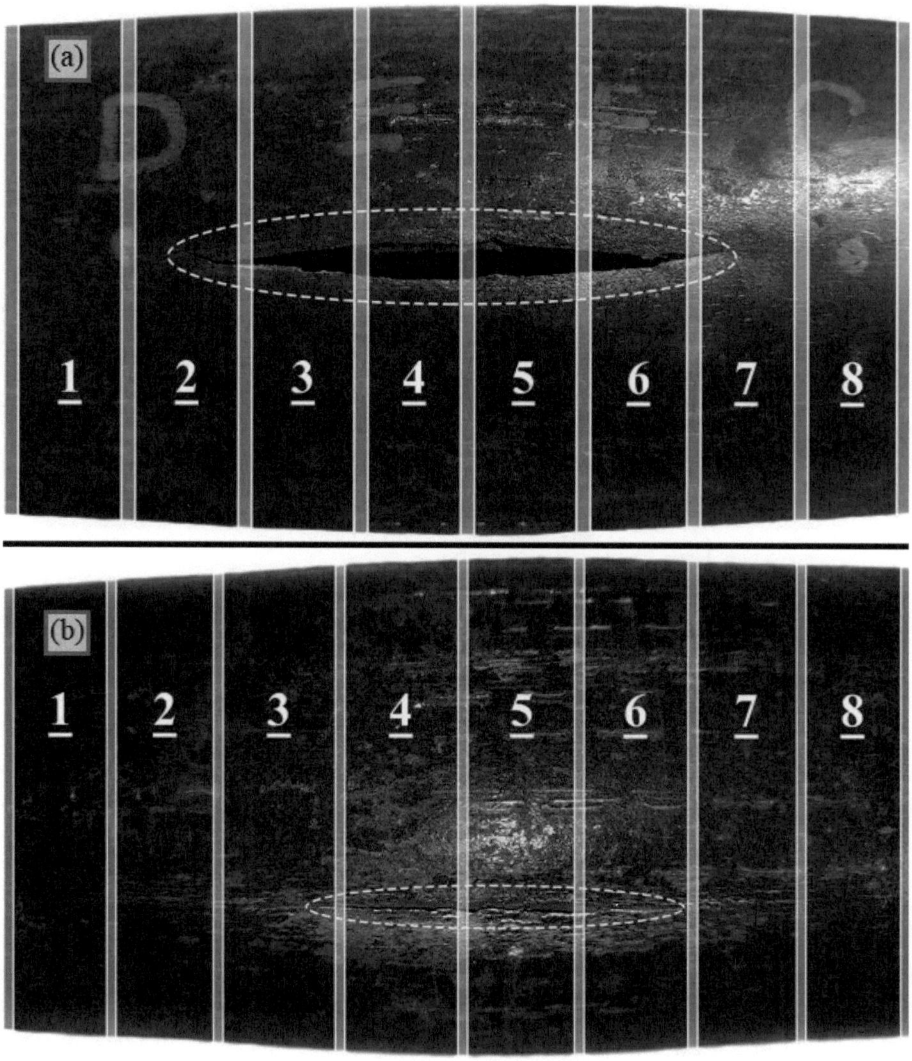

Figure 9.
Failure site of (a) A106-B and (b) A333-6 pipes due to overload during freezing test [25].

the pipe failure mode. The slant and double-slant fractures that can be observed in **Figure 10** are indicative of ductile failure due to plane stress loading conditions [38], which is a common form of failure for pressure vessels and pipes [39]. In these cases, the pipe failure begins with local thinning. Then localization of shear bands that occurs at 45°angle develops in the necked spot [40, 41]. The necked area that led to the rupture of the pipe in slant failure mode can be seen in **Figure 10**.

4.5 Fractography

The micrographs taken from the fracture surfaces confirm that the failure mode of both pipes was ductile tearing. This occurred because of the propagation of microcracks as a result of the coalescence of microvoids in the axial direction (along *z*-

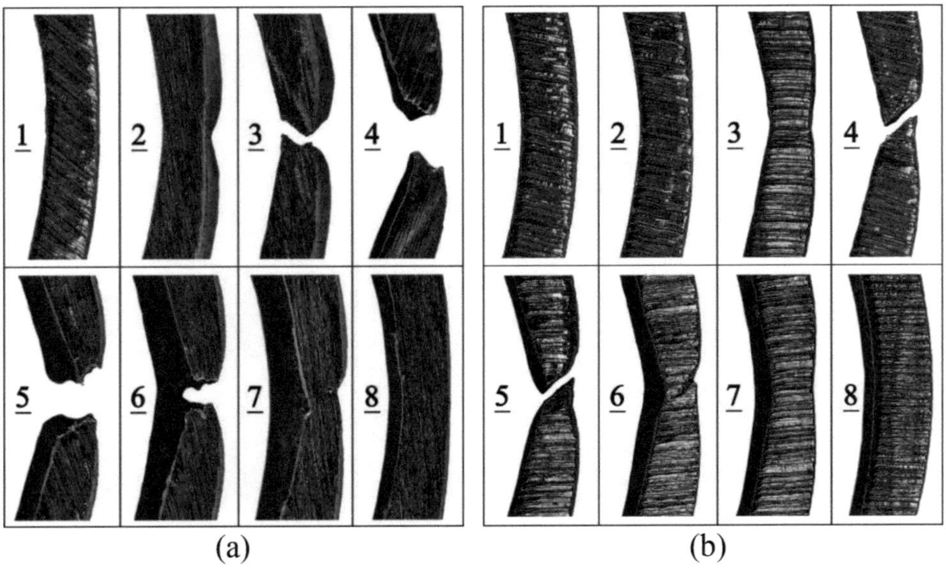

Figure 10.
Side views of the rings cut from (a) A106-B and (b) A333-6 pipes [25].

axis), as shown in **Figure 11**. Another feature that confirms the ductile failure mode of pipes is the fibrous and dull appearance of the fractographs that embodies numerous dimples.

4.6 Coating characterization

High-magnification images were taken in backscattered electron mode from the cross sections of the coated samples. These SEM images are shown in **Figure 12**. Although generation of microcracks was anticipated initially due to the brittle nature of the ceramic layer and thermal stress that can be generated at the interfaces due to the mismatch in material properties [42, 43], no major damage or delamination was observed during the microstructural evaluation of the coated samples even after conducting expedited deicing tests with high supplied power values.

The material for each layer of the coating system was selected based on the required mechanical and electrical properties to accomplish the given tasks. Alumina was selected due to its unique properties. While this material is a great electrical insulator, it conducts heat readily in comparison to other ceramic materials [26]. Therefore, this material was able to prevent short-circuiting between the pipe (substrate) and the heating element (nickel-chromium alloy), and at the same time, it was capable of transferring the generated heat by the heating element to the pipe to accomplish the deicing task [44, 45].

Although penetration of the nickel-chromium alloy inside the alumina layer was observed to some extent during the microstructural evaluation, this did not result in the malfunction of the coating-based heater. The penetration of the molten particles in the alumina layer was due to the presence of a network of open pores inside the ceramic coating, which is a characteristic of coatings fabricated by flame spraying process [45]. Given the effective thickness of alumina was reduced as a consequence

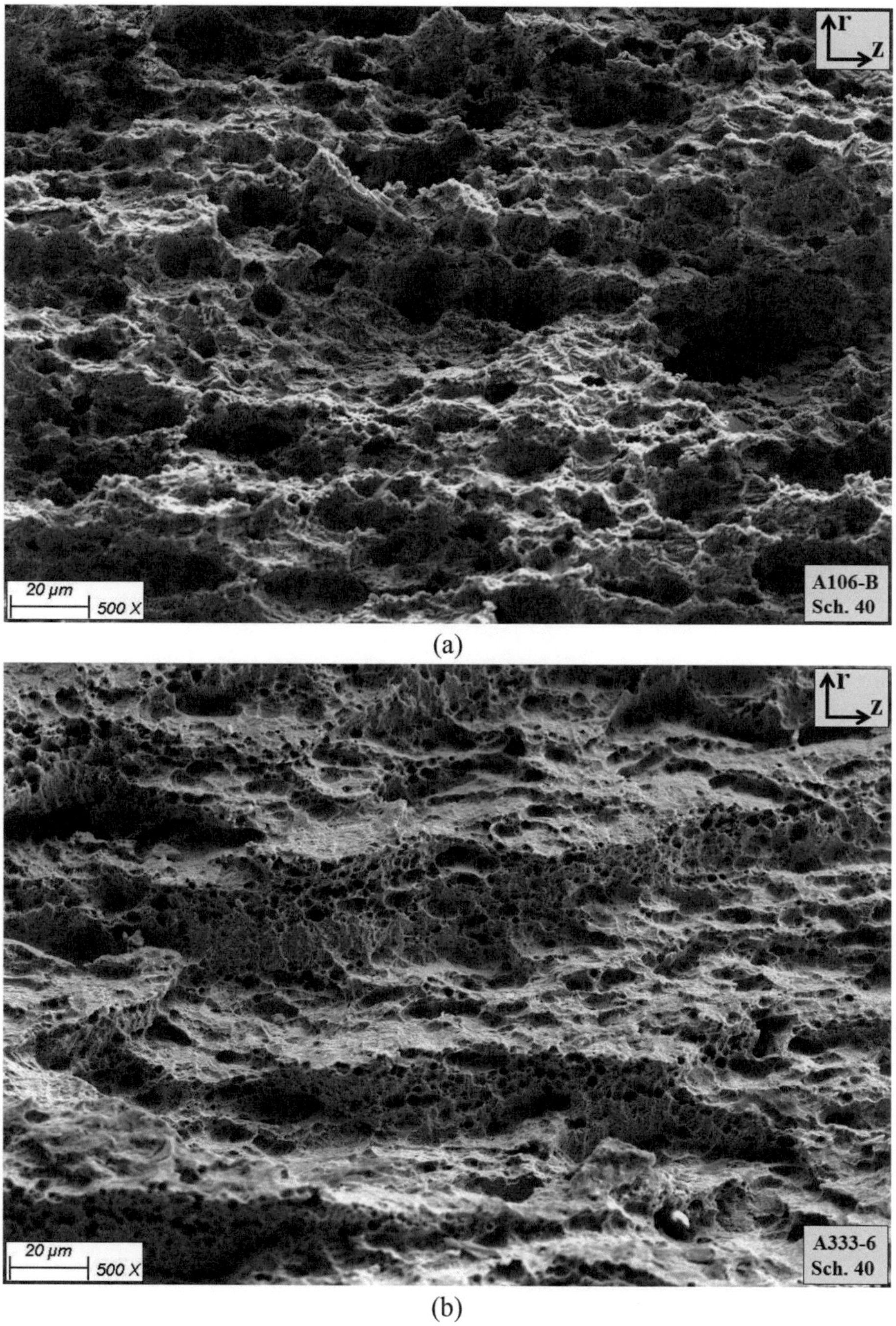

Figure 11.
Fractographs taken at 500X magnification from (a) A106-B pipe and (b) A333-6 pipe.

of this incident, alumina coatings with at least 50 microns thickness had to be deposited onto the pipe substrates to prevent any possible short-circuiting. It is well-established that much denser coatings can be fabricated by using more sophisticated

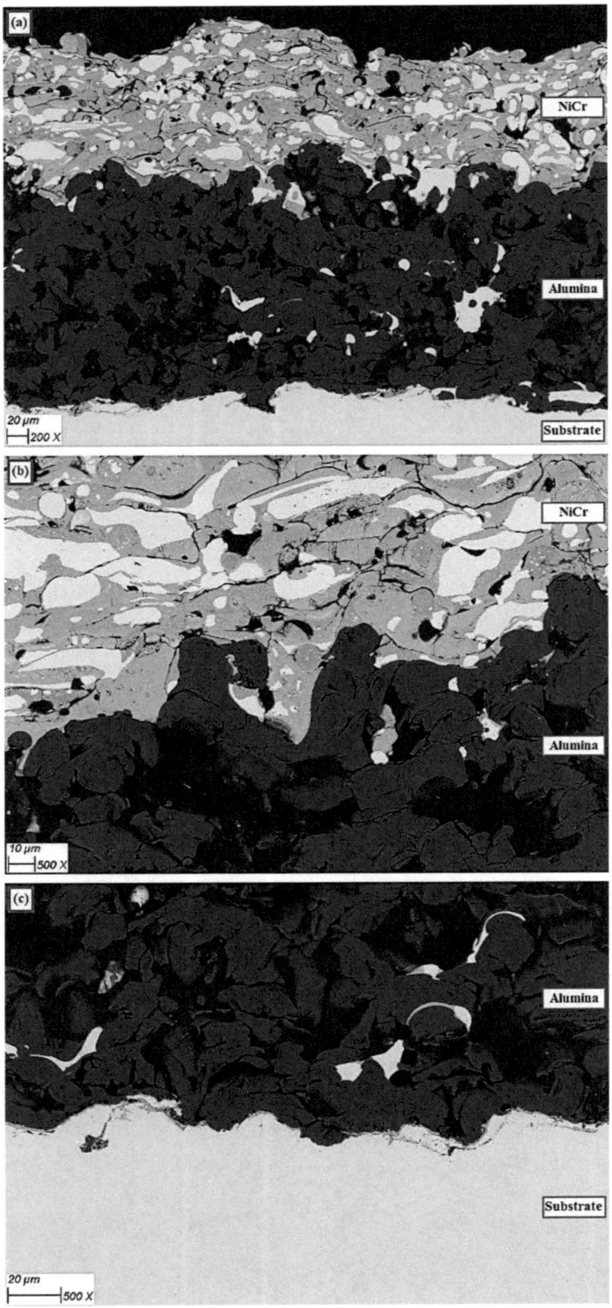

Figure 12.
SEM images taken from the coated pipe assembly: (a) bi-layered coating system, (b) NiCr-Al_2O_3 interface, and (c) Al_2O_3-steel interface [29].

thermal spray means such as plasma spraying, high-velocity oxygen fuel, or suspension plasma spray to prevent this undesirable occurrence [46, 47].

The nickel-chromium material was selected for the heating element because of its relatively high electrical resistivity. Although the electrical resistivity is a material property, the electrical resistance of the heating element is a function of its cross-

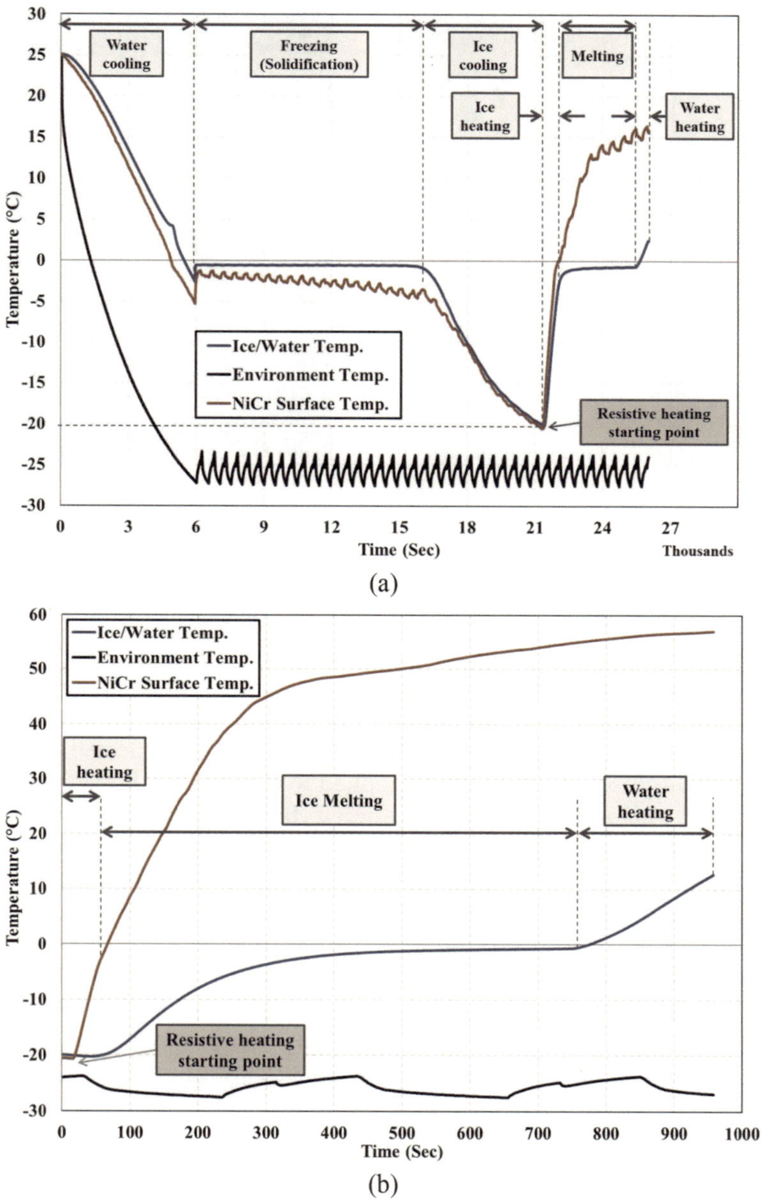

Figure 13.
Temperature traces from deicing tests for (a) free convection with supplied power of 80 W and (b) forced convection with supplied power of 500 W.

sectional area, which itself can be affected by discontinuities and presence of imperfections such as cracks and pores [45]. Presence of pores in the coating increases the electrical resistance as a result of reduction in the cross-sectional area through which free electrons can move [48]. It has been observed that formation and propagation of microcracks can also bring about the same consequences [49]. Therefore, the performance of the heating element that depends on its electrical resistance can be varied by using different thermal spraying methods.

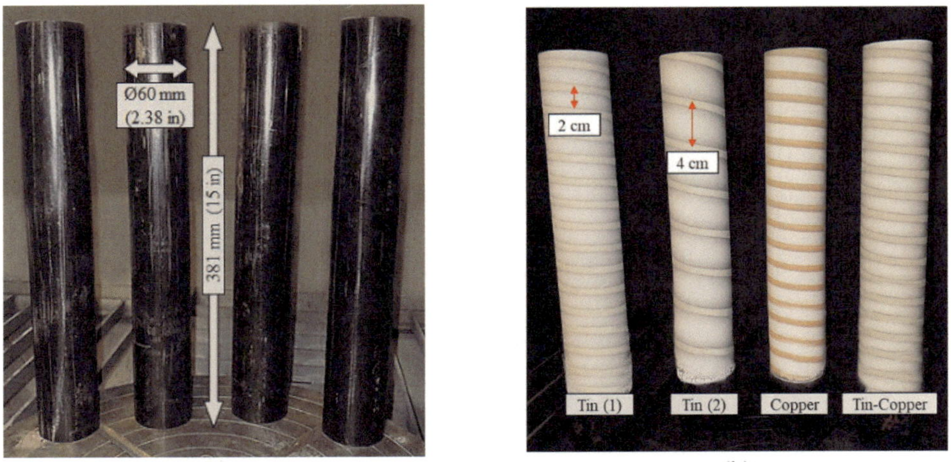

(a) (b)

Figure 14.
The presentation of (a) carbon steel pipes used as substrate and (b) pipes equipped with novel cold-sprayed tin and copper heating elements.

4.7 Heating performance

The performance of the coating-based heater was assessed under different power inputs. For this purpose, when 20 V was applied to the coating, 80 W power was generated, which was sufficient to heat and melt the ice inside the pipe. The temperature traces for the coating surface and enclosed ice/water temperatures can be seen in **Figure 13a** for the case of free convection. The functionality of the developed deicing system was evaluated based on the time that was required to accomplish the deicing test. It was observed that even under harsh conditions where the bare pipe was exposed to the forced convection due to the circulation of the cold ambient air, the fabricated coating-base heater melted the ice inside the pipe successfully as shown in **Figure 13b**. It was found that the voltage and current of the heating element were proportional and therefore, the nickel-chromium layer was an ohmic material [45].

4.8 Future work

In addition to the technical aspects, the economic and environmental implications of fabrication and utilization of the coating-based heating systems have also been investigated [50–52]. The promising results obtained from these studies confirmed the possibility of using the coating-based deicing systems on mass scale, however further advancement and improvement in the geometry of patterned coatings, manufacturing process, and the selection of the materials can even bring about more encouraging outcomes. Therefore, in the next phase of the project, in order to reduce the production cost, minimize the emission of exhaust gases to the environment, and achieve a more consistent and dense structure, the heating elements are fabricated by using cold spray system in helical pattern as shown in **Figure 14**. Other advantages of fabrication of heating element via cold spraying are prevention of penetration of the heating element inside the electrically insulating layer, on-site repairability, and enhanced bonding to the insulating layer thanks to the applied compressive stresses during the spraying process.

5. Conclusions

The freezing mechanism of the pressurized water inside steel pipes and the subsequent bulging and bursting were studied. In order to overcome this undesirable issue that brings about noticeable financial losses and environmental concerns, a novel deicing system was developed by using flame spraying and cold spraying processes. Once the performance of the developed heating system was assessed in simulated environmental conditions under different supplied powers, the coated pipes were sectioned so that their microstructure can be analyzed. The outstanding findings of this study are presented hereunder:

1. The bursting of the pipe is dependent on the pressurization of the enclosed water, which itself depends on the pipe size, material, and wall thickness. It is also a function of the prior work hardening of the pipe material, which is indicative of the number of times the pipe underwent freezing-thawing cycles.

2. The developed efficient coating-based heater was able to heat and melt the ice in the pipe easily even under harsh environmental conditions where bare pipes were exposed to low temperatures air under forced convection conditions.

3. It was found that the deicing performance of the coating heater was heavily dependent on the spraying method and spraying parameters.

4. The microstructural evaluation of the samples confirmed that the structural integrity of the coating system was not compromised even after conducting numerous freezing and heating cycles. The promising outcomes of this study emphasize the feasibility of using the coating-based heating system on mass scale for industrial applications.

Author details

Milad Rezvani Rad[1*] and Andre McDonald[2]

1 University of Southern Indiana, Evansville, Indiana, United States of America

2 University of Alberta, Edmonton, Alberta, Canada

*Address all correspondence to: milad.rad@usi.edu

References

[1] Smith KM, Van Bree MP, Grzetic JF. Analysis and testing of freezing phenomena in piping systems. In: Proceedings of the ASME 2008 International Mechanical Engineering Conference and Exposition. Boston, MA; 2008. pp. 123-128

[2] Edwards DB, Smith KM, Duvall DE, Grzetic JF. Analysis and testing of freezing phenomena in plastic piping systems. In: Proceedings of the ASME 2009 International Mechanical Engineering Conference and Exposition. Lake Buena Vista, Florida; 2009. pp. 201-206

[3] Carrns A. Avoiding costly home damage from Winter's cold. The New York Times. 2014

[4] Gordon JR. An Investigation into Freezing and Bursting Water Pipes in Residential Construction. Illinois: University of Illinois at Urbana-Champaign; 2006. pp. 1-51

[5] Sandberg C, King RM, Nguyen H, Foster MR. Analysis for Arctic valve heat tracing requirements. In: IEEE Transactions on Industry Applications. 2003;**39**(5):1462-1466. DOI: 10.1109/TIA.2003.816501

[6] Deshmukh YV. Industrial Heating: Principles, Techniques, Materials, Applications, and Design. Boca Raton: Taylor & Francis; 2005. p. 358

[7] Sandberg C et al. Panel discussion on electrical heat tracing. In: 1999 IEEE Power Engineering Society Summer Meeting. Conference Proceedings. Vol. 2. Edmonton, Alta., Canada; 1999. pp. 900-903

[8] Erickson CJ. Reliable and cost-effective electrical heating of pipelines with self-regulating heating cables. IEEE Transactions on Industry Applications. 1988;**24**(6):1089-1095

[9] Johnson BC, Propst JE. Technological advancement in self-limiting electric heat tracing cables. In: Industrial Applications Society 35th Annual Petroleum and Chemical Industry Conference. Dallas, TX; 1988. pp. 9-13

[10] Michels D, Hadeler J, Lienhard J. High-heat-flux resistance heaters from VPS and HVOF thermal spraying. Experimental Heat Transfer. 1998;**11**:341-359

[11] Killinger A, Gadow R. Thermally sprayed coating composites for film heating devices. Advances in Science and Technology. 2006;**45**:1230-1239

[12] Prudenziati M. Development and the implementation of high-temperature reliable heaters in plasma spray technology. Journal of Thermal Spray Technology. 2008;**17**(2):234-243

[13] Floristan M, Fontarnau R, Killinger A, Gadow R. Development of electrically conductive plasma sprayed coatings on glass ceramic substrates. Surface and Coating Technology. 2010;**205**(4):1021-1028

[14] Lamarre J et al. Performance analysis and modeling of thermally sprayed resistive heaters. Journal of Thermal Spray Technology. 2013;**22**(6):947-953

[15] Lopera-Valle A, McDonald A. Application of flame-sprayed coatings as heating elements for polymer-based composite structures. Journal of Thermal Spray Technology. 2015;**24**(7):1289-1301

[16] Sampath S et al. Thermal spray techniques for fabrication of Meso-electronics and sensors. MRS Proceedings. 2000;**624**:181

[17] Herman H, Sampath S, McCune R. Thermal spray: Current status and future trends. MRS Bulletin. 2000;**25**(7): 17-25

[18] Pawłowski L. The relationship between structure and dielectric properties in plasma-sprayed alumina coatings. Surface and Coatings Technology. 1988;**35**(3):285-298

[19] Luo R et al. Structure and electrical insulation characteristics of plasma-sprayed alumina coatings under pressure. Ceramics International. 2018; **44**(6):6033-6036

[20] Toma F et al. Comparative study of the electrical properties and characteristics of thermally sprayed alumina and spinel coatings. Journal of Thermal Spray Technology. 2011;**20**(1): 195-204

[21] Yan B-D, Meilink S, Warren G, Wynblatt P. Water Adsorption and Surface Conductivity Measurements on alpha-Alumina Substrates. In: IEEE Transactions on Components, Hybrids, and Manufacturing Technology. 1987;**10** (2):247-251. DOI: 10.1109/TCHMT.1987.1134727

[22] Toma F et al. Comparison of the microstructural characteristics and electrical properties of thermally sprayed Al_2O_3 coatings from aqueous suspensions and feedstock powders. Journal of Thermal Spray Technology. 2012;**21**(3):480-488

[23] Stahr CC et al. Dependence of the stabilization of α-alumina on the spray process. Journal of Thermal Spray Technology. 2007;**16**(5):822-830

[24] Younis HF, Dahbura RS, Lienhard JH. Thin film resistance heaters for high heat flux Jet-Array cooling experiments. Proceedings of the ASME 1997 international mechanical engineering congress and exposition. Heat Transfer: Volume 3. Dallas, Texas, USA: ASME; 1997. pp. 127-134

[25] Rezvani Rad M, Marsden R, McDonald A. Testing analysis of freezing phenomenon in conventional carbon steel pipes. In: Proceedings of the Canadian Society for Mechanical Engineering International Congress 2018, May 27–30, 2018. Toronto, ON, Canada: The Canadian Society for Mechanical Engineering; 2018

[26] Metco O. Material product data sheet: DSMTS-0005.6—Aluminum oxide thermal spray powders. Oerlikon Metco, Westbury. 2017;**2017**:1

[27] Praxair Surface Technologies Inc. Powder Solutions Catalog. Indianapolis: Praxair Surface Technologies Inc; 2014. p. 12

[28] Centerline (Windsor) Limited. Technical Data Sheet: SST-TDS-C5003-PR-1.3-0415—Commercial Copper Powder. Windsor, Ontario: Centerline (Windsor) Limited; 2017. p. 1

[29] Rezvani Rad M, McDonald A. Development of a flame-sprayed coating system to mitigate ice accumulation and freezing damage in carbon steel pipes. In: International Thermal Spray Conference, May 7–10, 2018. Orlando, FL, USA: ASM International; 2018

[30] Rezvani Rad M, McDonald A. Mathematical simulation of heating and melting of solid ice in a carbon steel pipe coated with a resistive heating system. International Journal of Heat and Mass Transfer. 2019;**138**: 923-940

[31] McDonald A, Bschaden B, Sullivan E, Marsden R. Mathematical simulation of the freezing time of water in small diameter pipes. Applied Thermal Engineering. 2014;**73**(1):142-153

[32] Gilpin RR. The effects of dendritic ice formation in water pipes. International Journal of Heat and Mass Transfer. 1977;**20**:693-699

[33] Gilpin RR. The influence of natural convention on dendritic ice growth. Journal of Crystal Growth. 1976;**36**: 101-108

[34] Choukroun M, Grasset O. Thermodynamic model for water and high-pressure ices up to 2.2GPa and down to the metase domain. The Journal of Chemical Physics. 2007;**127**:124506

[35] ASTM A106/A106M-13. Standard Specification for Seamless Carbon Steel Pipe for High-Temperature Service. West Conshohocken, PA: ASTM International; 2013. p. 3

[36] ASTM A333/A333M-11. Standard Specification for Seamless and Welded Steel Pipe for Low-Temperature Service. West Conshohocken, PA: ASTM International; 2011. p. 3

[37] PVRC. In: Welding Research Council, editor. Recommendations on Toughness Requirements for Ferritic Materials. 1972

[38] Becker WT. Fracture Appearance and Mechanisms of Deformation and Fracture, Failure Analysis and Prevention. Materials Park, Ohio: ASM International; 2002. pp. 1162-1327

[39] Huang H, Xue L. Prediction of slant ductile fracture using damage plasticity theory. International Journal of Pressure Vessels and Piping. 2009;**86**:319-328

[40] El-Naaman SA, Nielsen KL. Observations on mode I ductile tearing in sheet metals. European Journal of Mechanics—A/Solids. 2013; **42**:54-62

[41] Nielsen KL, Hutchinson JW. Cohesive traction–separation laws for tearing of ductile metal plates. International Journal of Impact Engineering. 2012;**48**:15-23

[42] Rezvani Rad M, Farrahi GH, Azadi M, Ghodrati M. Effects of preheating temperature and cooling rate on two-step residual stress in thermal barrier coatings considering real roughness and porosity effect. Ceramics International. 2014;**40**:15925-15940

[43] Rezvani Rad M, Farrahi GH, Azadi M, Ghodrati M. Stress analysis of thermal barrier coating system subjected to out-of-phase thermo-mechanical loadings considering roughness and porosity effect. Surface and Coatings Technology. 2015;**262**:77-86

[44] Rezvani Rad M, Mohammadian Bajgiran M, Moreau C, McDonald A. Microstructural and performance analyses of thermal-sprayed electric resistance heating systems deposited onto steel pipes as de-icing elements. In: International Thermal Spray Conference, May 26–29, 2019. Yokohama, Japan: ASM International; 2019. p. 8

[45] Rezvani Rad M, Mohammadian Bajgiran M, Moreau C, McDonald A. Fabrication of thermally sprayed coating Systems for Mitigation of ice accumulation in carbon steel pipes and prevention of pipe bursting. Surface and Coatings Technology. 2020;**397**: 126013

[46] Mohammadian Bajgiran M, Rezvani Rad M, McDonald A, Moreau C.

Microstructure, phase and dielectric strength of thermally sprayed alumina layers in coating–based heating systems. International Journal of Applied Ceramic Technology. 2021;**00**:1-16

[47] Mohammadian Bajgiran M, Rezvani Rad M, McDonald A, Moreau C. On reliability of thermally- sprayed alumina dielectric layers in a multi-layered coating-based heating system. In: Proceedings of the Canadian Society for Mechanical Engineering International Congress. London; 2019. p. 7

[48] Ashrafizadeh H et al. Deposition of electrically conductive coatings on castable polyurethane elastomers by the flame spraying process. Journal of Thermal Spray Technology. 2016;**25**(3): 419-430

[49] Bobzin K et al. Temperature distribution on thermally sprayed heating conductor coatings. IOP Conference Series. 2019;**480**:012002

[50] Rezvani Rad M, Ngaokere K, Hayden D, Kumar A, McDonald A. Techno-economic assessment of coating-based resistive heating systems versus conventional heat tracers. In: International Thermal Spray Conference, May 26–29, 2019. Yokohama, Japan: ASM International; 2019. p. 7

[51] Rezvani Rad M, Ngaokere K, Lloyd S, McDonald A. Techno-economic assessment of multi-layered coating-based resistive heating systems and conventional heat tracing cables for industry pipes. Journal of Thermal Sprayed Technology. 2021;**30**:1393-1409

[52] Yousefzadeh Z, Rezvani Rad M, McDonald A, Lloyd S. Life Cycle Assessment of a Thermal Sprayed Al2O3-NiCr Resistive Heating Coating for Pipe Freeze Protection. Journal of Thermal Sprayed Technology. 2022;**31**:378-395

Chapter 3

Evaluation of Electromagnetic Interferences Affecting Metallic Pipelines

Denisa Șteț, Levente Czumbil and Dan Doru Micu

Abstract

This chapter presents some analysis of the modeling techniques used to evaluate the effects of electromagnetic interference phenomena that could occur when metallic pipelines are placed close to high-voltage power lines. The electric and magnetic fields produced by overhead power lines could perturb the normal operation of the metallic pipelines through induced currents and voltages. These perturbations could be dangerous for both pipeline operating personnel (as electrical hazard) and pipeline structural integrity (due to accelerated electrochemical corrosion phenomena). The chapter depicts the electromagnetic coupling mechanisms behind the abovementioned interference phenomena and how the induced voltages could be evaluated. A parametric analysis is showcased to highlight the influence of various geometrical and electrical parameters.

Keywords: metallic pipelines, electromagnetic interferences, power lines influence, modeling and simulation, electromagnetic compatibility, induced voltages

1. Introduction

European regulations regarding environmental protection, as well as economic reasons aimed to reduce construction costs, have determined a significant decrease in the access of energy carrier (like oil products, electricity, methane gas, or/and water) infrastructure to new right of ways. Therefore, the transport and distribution network of water, natural gas, or crude oil pipelines must share on long distances (for several kilometers) the same distribution corridors with high-voltage power lines (HVPL) (see **Figure 1**).

Studies on high-voltage AC power lines operating in steady state or fault regime identified the presence of electromagnetic interferences in nearby metallic structures due to the electric and magnetic fields generated by the currents flowing through the HVPL conductors [1–4].

Therefore, in many situations, the metallic pipeline (MP) used for the transport and distribution of liquid or gaseous substances may be exposed to induced currents or voltages. These could be dangerous on one side for operating personnel, who coming into contact with the pipeline may be exposed to electrocution, and on the

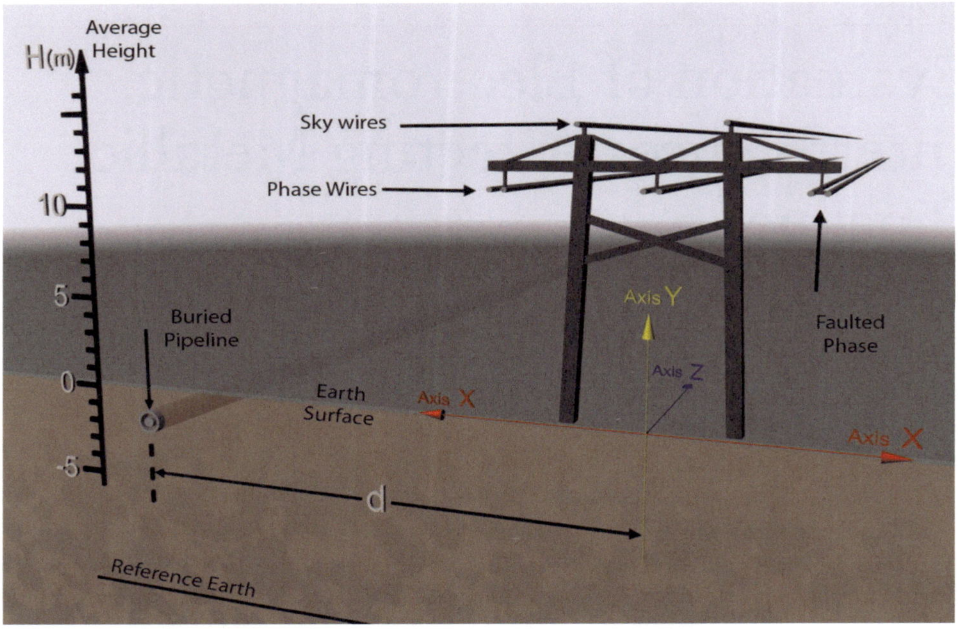

Figure 1.
Cross section of a common HVPL-MP right-of-way.

other hand, to the structural integrity of the pipeline due to accelerated corrosion phenomena [5–7].

In 1995, within the CIGRE Working Group 36.02, the document entitled Guide Concerning Influence of High Voltage AC Power Systems on Metallic Pipelines [8] has been developed, which addresses the influence of high-voltage energy devices on nearby metallic pipelines. It is a reference document in the field of electromagnetic interferences and summarizes the following aspects:

- presentation of different categories of electromagnetic disturbances and the problems arising from them

- description of simple methods for evaluating electromagnetic interference phenomena and their measurement methods

- presentation of main ways and means of reducing their influences, as well as the description of the most important protection systems.

Regarding the protection of underground MP against electrochemical corrosion processes, a special attention has been paid to the effects caused by AC systems. Previously, AC corrosion was considered negligible compared to the interferences in DC state. But research in the field proved the opposite, resulting in pipeline protection guidelines [9, 10] and a European regulatory standard [11]. The main causes of AC corrosion are the leakage current densities from the pipeline into the surrounding soil. The source of leakage currents is the AC voltages induced in the pipelines due to various electromagnetic coupling mechanisms [6, 7].

To comply with the European norms and regulations, it is necessary to investigate the electromagnetic interference phenomena in detail and to precisely determine the magnitude of the voltages induced in MPs exposed to nearby power lines, both under normal and HVPL fault operating conditions. To identify and apply the proper protection techniques for the metallic pipelines, the geometrical and electrical parameters that influence the level of induced voltages must be determined. The difficulty in accessing buried MPs not only requires mathematical models for the numerical evaluation of induced voltages, instead of on-site measurements [12].

2. Electromagnetic interference phenomena

At low frequencies, when the wavelength of the electrical signal is long compared to the size of the perturbation source, the electromagnetic interference propagates through capacitive (electric), inductive (magnetic), or conductive (galvanic) couplings [8, 11, 13]. Conversely, at high frequencies when the wavelength is comparable to the dimensions of the perturbation source, electromagnetic radiation appears. The radiation transition limit is variable, but in most practical cases, it is approximately 10 m, which corresponds to a frequency of 30 MHz. Therefore, in case of interference problems between HVPL and neighboring metallic structures, only conductive, capacitive, and inductive couplings should be considered [8, 11].

2.1 Galvanic coupling

Galvanic (conductive) couplings occur when two electrical circuits have a common portion of the circuit. In this situation, electromagnetic interference propagates in the receiver (victim) through conduction, through one or more conductors (power line, cable shield, etc.), or even through passive elements (capacitors, transformers, etc.).

The energy transported through power lines can be transmitted conductively to adjacent metallic structures (underground metallic pipes, etc.) if these structures are connected to the AC circuit (the grounding grid of a power line tower or any other device that has a network extended earthing), either directly (metallic connection) or through the proximity of the pipeline and the grounding grid (see **Figure 2**).

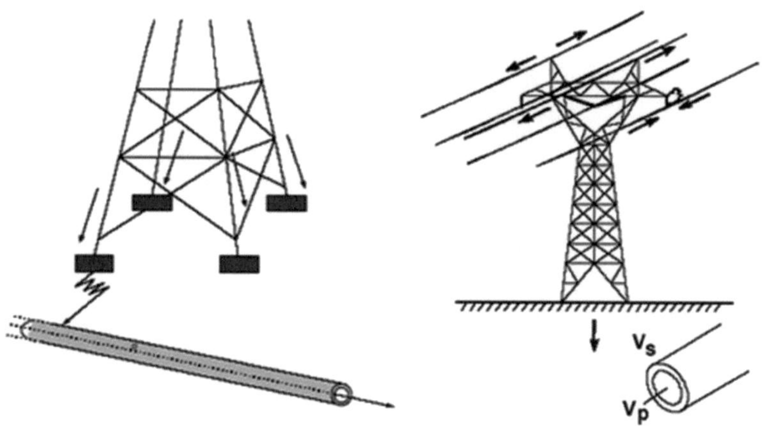

Figure 2.
Galvanic coupling between a HVPL and a MP.

The occurrence of a ground fault in the power system corresponds to a high-value short-circuit current flowing through earth. The electrical potential of the soil (close to the grounding grid) will increase considerably in relation with the pipeline potential (assumed to be at a reference potential) [14]. Any element connected between MP and the ground will be subject to this potential difference; in other words, the fault currents that flow through the earthing electrode of an HVPL tower, or substation, produce an increase in the potential of the electrode as well as the potential of the soil in its vicinity compared to the reference potential. Under these conditions, the metallic structures (pipelines) will be influenced if they are directly connected to the grounding grid of the electric energy transport/distribution system (for example, of a transformer station), or if they cross the influence zone near the grounding sockets. In these cases, the potential of the pipeline increases and can be transmitted over a relatively large distance (several km), depending on the degree of electrical insulation of the interfered metallic structure and the resistivity of the soil in the respective area.

2.2 Capacitive coupling

Capacitive (electric) coupling occurs between two circuits whose conductors are at different potentials. As a result of the potential difference between the conductors, an electric field is produced and modeled in an equivalent electrical scheme by a parasitic capacitance. **Figure 3** presents an example of electrical coupling of two circuits (1) and (2) by means of a quasi-stationary electric field, respectively, through parasitic capacities.

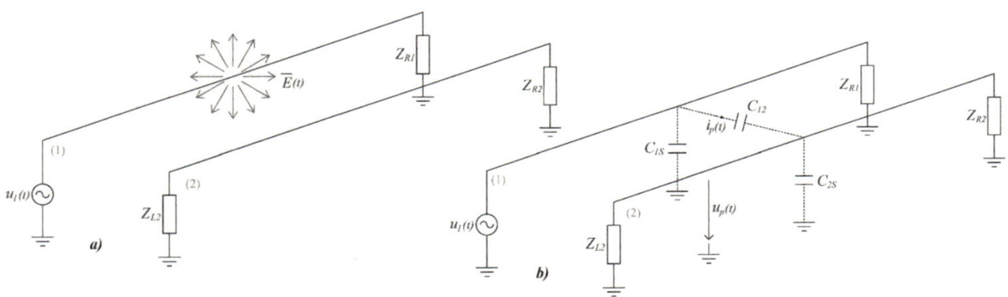

Figure 3.
(a) Field model, and (b) network model for capacitive coupling.

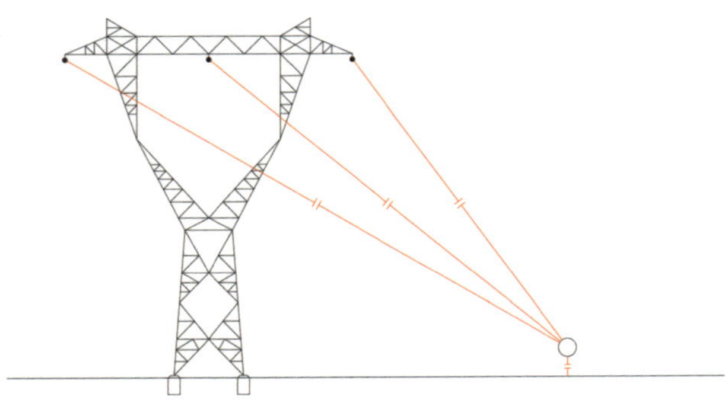

Figure 4.
Capacitive coupling between HVPL and metallic pipeline.

In the case of capacitive coupling, determined by the electric field of the high-voltage power line, the hypothesis that the soil is a perfect conductor, of infinite conductivity, can be used, because its conductivity is much higher than that of air (see **Figure 4**).

Therefore, only aboveground metallic structures located in the vicinity of HVPL are subject to capacitive coupling perturbations; this effect appears both in normal operating conditions and power line fault conditions [15].

2.3 Inductive coupling

Inductive (magnetic) couplings occur between two or more circuits passed by electric currents producing time-varying magnetic fields that induce in the victim circuit a voltage that disturbs the desired signal (if exists). The action of the magnetic field of the disturbing circuit on the perturbed one can be represented in the equivalent electric circuit model by a mutual inductance or by an induced voltage source (see **Figure 5**).

A metallic pipeline near a high-voltage power line (as shown in **Figure 6**) is subject to induced voltages by magnetic coupling. In other words, a significant part of the energy transmitted through power lines surrounds the conductors and extends over large distances from their center. The nearby metallic pipelines, which have a common path with HVPL, can capture this energy in unfavorable parallelism and/or power line operation conditions (high

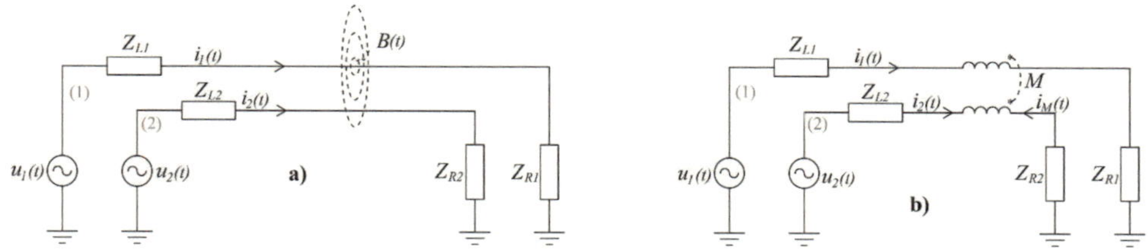

Figure 5.
(a) Field model, and (b) network model for inductive coupling.

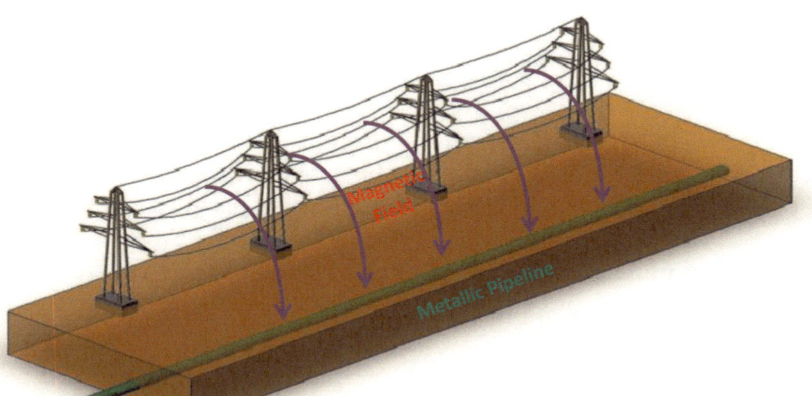

Figure 6.
Inductive coupling between HVPL and metallic pipeline.

zero-sequence currents, phase to ground faults, unbalanced loads, non-symmetric current system, etc.)

As shown in some previous works, such as [16–18], the inductive influence from 50 Hz or 60 Hz electric power systems on pipeline networks is related to the time-varying magnetic field generated by the electric currents flowing in power lines conductors. The main parameters generally considered for this type of influence are the electric power system's current load level, the ground wire current, the exposure length, the separation distance between involved structures of both systems, and the soil model resistivity [19, 20].

Given that the pipeline is buried and assuming that the ground permittivity can be neglected, there is no capacitive influence between the overhead transmission line and the buried pipeline. In this regard, the next chapters emphasize aspects related to investigation of the induced AC voltage effects on a buried pipeline due to the currents in an overhead transmission line.

3. Modeling techniques of the electromagnetic interference problems

Numerous numerical simulations have been carried out over the past few decades to analyze interference problems involving power lines and earth-return conductors. There are two main approaches for assessing such phenomena: transmission line theory [1, 21–23] and a hybrid approach based on finite element method (FEM) and circuit analysis [2, 3, 24–26]. The first approach has the advantage that it can be used for any operation state, while the other is only applicable to steady-state conditions.

In either case, a fundamental requirement of such modeling technique is that they should consider soil resistivity variations [24, 27, 28], non-parallel pipeline-power line right of ways, and multiple metallic conductors, such as sky wires and/or mitigation wires. Various commercial computer software are available to assess the influence of electric power systems on pipeline systems, but their licensing and annual renewal fees could be very high.

The analytical forms of the solutions of the electromagnetic field problems, related to the cylindrical current-carrying conductors in the presence of the ground, are a rather laborious problem, even when the geometry of the conductor network is simple. There are several difficulties with buried pipelines:

- Underground MPs can be long or short, but they cannot be infinitesimal or infinitely long.
- MPs are generally insulated and are in direct contact with the ground, so longitudinal (axial) and transversal (leakage) currents must be considered.
- Conductors with semi-insulating coating, such as pipelines, must also be analyzed.
- The most important results related to underground pipeline networks are related to the near field generated by these conductors.

In many cases [3, 15, 19, 20, 26, 28], the pipeline network is analyzed for low-frequency energizing currents conditions, but problems could arise when the performance of the network at high frequency should be investigated and transient state studies are required [18, 21, 23]. For both situations, an equivalent electrical circuit model of the investigated interference problem (like the one in **Figure** 7) must be constructed and analyzed considering all the present metallic conductors and all the coupling mechanisms.

By solving the equivalent electric circuit model of the entire right-of-way for the proper HVPL operating conditions, the induced currents and voltages could be evaluated along the pipeline length, and it could be determined if the pipeline would operate normally or mitigation measures should be applied.

3.1 Transmission line approach

The transmission line approach is the most general and complex formulation of an HVPL-MP interference problem that could be applied for both steady state and transient state studies. It has been developed from the well-known lumped-element model for an infinitesimal (Δx length) single-wire conductor presented in **Figure 8**.

Imposing $\Delta x \rightarrow 0$ the following telegrapher's equations can be written:

$$-\frac{\partial u(x,t)}{\partial x} = R_0 \cdot i(x,t) + L_0 \cdot \frac{\partial i(x,t)}{\partial t} \tag{1}$$

$$-\frac{\partial i(x,t)}{\partial x} = G_0 \cdot u(x,t) + C_0 \cdot \frac{\partial u(x,t)}{\partial t} \tag{2}$$

where R_0 is per-unit length resistance, L_0 per-unit length inductance, G_0 per-unit length conductance, and C_0 per-unit length capacitance. If Laplace transformation ($s = j\omega$) is applied to the above equations, we will obtain:

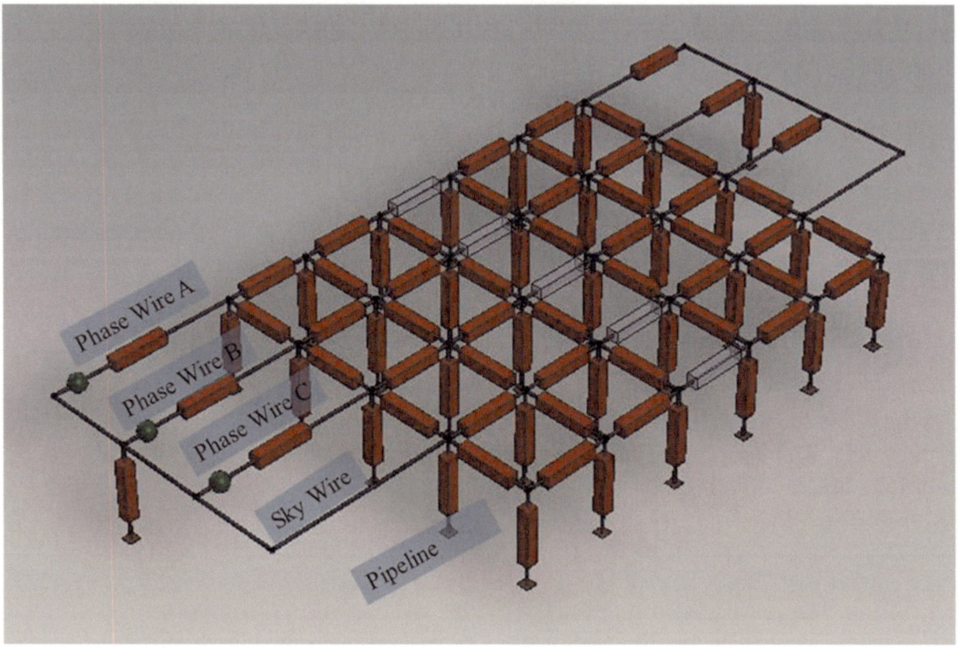

Figure 7.
Equivalent circuit model for a general case of HVPL–MP interference.

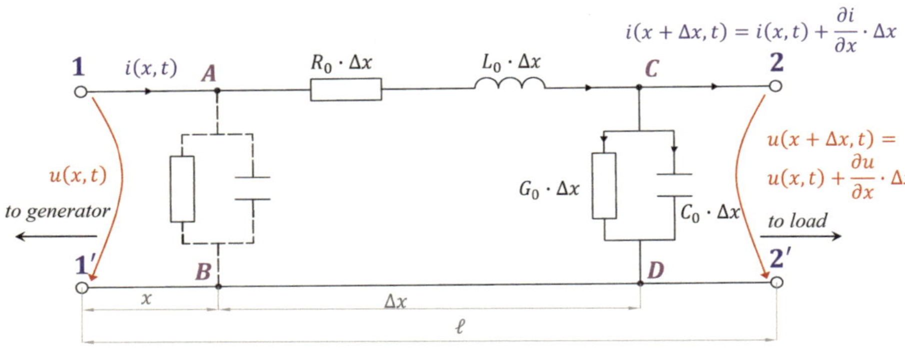

Figure 8.
Lumped-element equivalent circuit transmission line model.

$$-\frac{\partial u(x,s)}{\partial x} = (R_0 + sL_0)\cdot i(x,s) = Z_0 \cdot i(x,s) \tag{3}$$

$$-\frac{\partial i(x,s)}{\partial x} = (G_0 + sC_0)\cdot u(x,s) = Z_0 = Y_0 \cdot u(x,s) \tag{4}$$

where Z_0 is the longitudinal impedance, and Y_0 is the shunt admittance
The solution of Eqs. (3) and (4) can be written as:

$$u(x,s) = e^{-\gamma\cdot x}\cdot u^+ + e^{\gamma\cdot x}\cdot u^- \tag{5}$$

$$i(x,s) = \frac{1}{Z_C}\left(e^{-\gamma\cdot x}\cdot u^+ - e^{\gamma\cdot x}\cdot u^-\right) \tag{6}$$

where u^+ and u^- denote the voltage as incident and reflect waves, respectively, γ is the propagation constant, and Z_C is the characteristic impedance, defined as:

$$\gamma = \sqrt{Z_0\cdot Y_0} \tag{7}$$

$$Z_C = \sqrt{\frac{Z_0}{Y_0}} \tag{8}$$

The substitution of boundary conditions $x = 0$ and $x = l$ where l denotes the length, into (5) and (6) leads to the following line terminals equations:

$$u_k - Z_C\cdot i_k = e^{-\gamma\cdot x}\cdot(u_m + Z_C\cdot i_m) \tag{9}$$

$$u_k + Z_C\cdot i_k = e^{\gamma\cdot x}\cdot(u_m - Z_C\cdot i_m) \tag{10}$$

where k and m denote the terminals of the line corresponding to $x = 0$ and $x = l$, respectively. The currents i_k and i_m are entering the line at both ends. The above equations are rewritten to:

$$i_k = Y_C\cdot u_k - H(Y_C\cdot u_m + i_m) \tag{11}$$

$$i_m = Y_C\cdot u_m - H(Y_C\cdot u_k + i_k) \tag{12}$$

where $Y_C = 1/Z_C$ is the characteristic admittance, and $H = e^{\gamma\cdot l}$ is the propagation function. Eqs. (11) and (12) can be rewritten using vector and matrices for a multi-conductor system:

$$[\boldsymbol{I_k}] = [\boldsymbol{Y_C}] \cdot [\boldsymbol{U_k}] - H([\boldsymbol{Y_C}] \cdot [\boldsymbol{U_k}] + [\boldsymbol{I_m}]) \tag{13}$$

$$[\boldsymbol{I_m}] = [\boldsymbol{Y_C}] \cdot [\boldsymbol{U_m}] - H([\boldsymbol{Y_C}] \cdot [\boldsymbol{U_k}] + [\boldsymbol{I_k}]) \tag{14}$$

To determine the frequency-dependent characteristic impedances and admittances required for transient state analysis, **Figure 9** illustrates the transition from a linear earth electrode to an equivalent transmission line with complex values, for the frequency-dependent parameters $\underline{Z}_C(\omega)$ and $\underline{\gamma}(\omega)$, which denote the characteristic impedance and propagation function, respectively:

$$\underline{Z}_C(\omega) = \sqrt{\frac{R_0 + j\omega L_0}{G_0 + j\omega C_0}} = \sqrt{\frac{\underline{Z}_0(\omega)}{\underline{Y}_0(\omega)}}, \underline{\gamma}(\omega) = \sqrt{\underline{Z}_0(\omega) \cdot \underline{Y}_0(\omega)} \tag{15}$$

$$\underline{\Gamma}_0(\omega) = j\omega\sqrt{\mu_0 \cdot \varepsilon_0} \tag{16}$$

This form corresponds to any two-conductor transmission line. Sunde derived equivalent expressions for a single conductor in contact to the soil, with a current returning through the earth [29, 30]. Both the complex longitudinal impedance per unit length, $\underline{Z}_0$, and the complex transversal admittance per unit length $\underline{Y}_0$, of a horizontal conductor consist of an internal term and an earth return term in the following manner:

$$\underline{Z}_0 \cong \underline{Z}_{0i} + \frac{j\omega\mu_0}{2\pi} \cdot ln \frac{1.85}{\sqrt{\underline{\gamma}^2 + \underline{\Gamma}^2} \cdot \sqrt{2ah}} \tag{17}$$

$$(\underline{Y}_0)^{-1} \cong (\underline{Y}_{0i})^{-1} + \frac{j\omega\mu_0}{\pi \cdot \underline{\Gamma}^2} \cdot ln \frac{1.12}{\underline{\gamma} \cdot \sqrt{2ah}} \tag{18}$$

Here $\underline{Z}_{0i}$ denotes the internal complex impedance of a conductor of radius a, buried at depth h, and $\underline{Y}_{0i}$ stands for the insulation admittance of an eventual coating gap, through which the conductor is in contact to the surrounding medium. The latter is characterized by the earth conductivity σ_E, relative electric permittivity ε_{rE}, and magnetic permeability μ_0. All three lead to a propagation function [31–33]:

$$\underline{\Gamma}(\omega) = \sqrt{j\omega\mu_0 \cdot (\sigma_E + j\omega\varepsilon_0\varepsilon_{rE})} \tag{19}$$

which would govern the transmission of impulses along the conductor if it were imbedded in a homogeneous soil with these parameters.

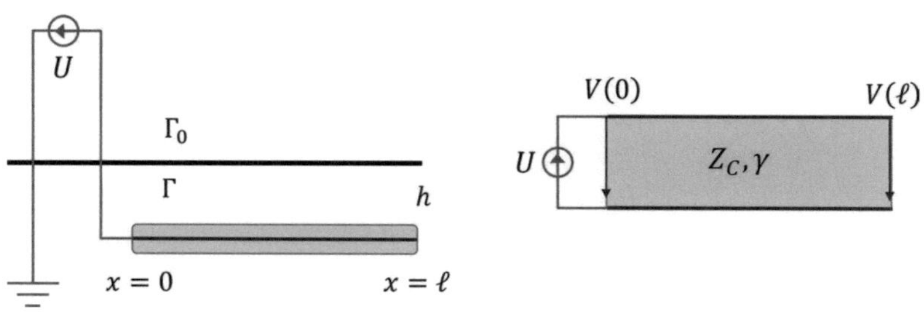

Figure 9.
Characteristic impedance of a linear earth electrode.

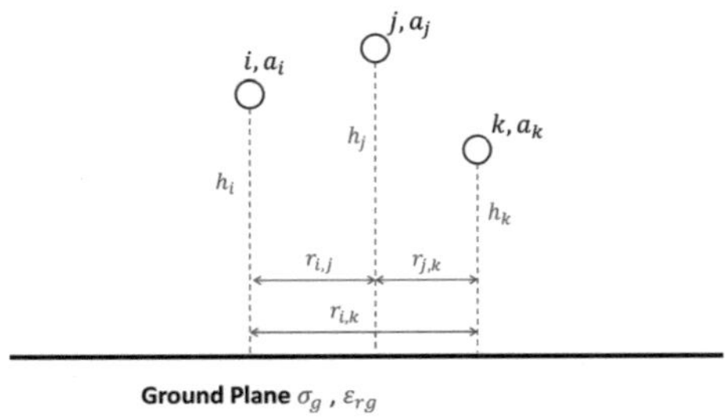

Figure 10.
Multiple conductors over a conductive earth.

For a more complex geometry like the one from **Figure 10**, the generalized telegrapher's equations for the case of multi-wire system along the x-axis above an imperfectly conducting ground and in the presence of an external electromagnetic excitation are [33]:

$$\begin{cases} \frac{d}{dx}[\underline{U}_i(x)] + [j\omega L_{ij}] \cdot [\underline{I}_i(x)] + [\underline{Z}_{gij}] \cdot [\underline{I}_i(x)] = [\underline{S}_{1i}^e(x)] \\ \frac{d}{dx}[\underline{I}_i(x)] + [G_{ij}] \cdot [\underline{U}_i(x)] + [j\omega C_{ij}] \cdot [\underline{U}_i^s(x)] = [\underline{S}_{2i}^e(x)] \end{cases} \tag{20}$$

where:

- $[\underline{U}_i(x)]$, $[\underline{I}_i(x)]$ are vectors of voltage and current along the line, in the frequency domain
- $[L_{ij}(x)]$, $[G_{ij}(x)]$, and $[C_{ij}(x)]$ are the matrices of per unit length line inductance, transverse conductance, and capacitance, respectively
- $[\underline{Z}_{gij}]$ is the matrix of ground impedance
- $[\underline{S}_{1i}^e(x)]$ and $[\underline{S}_{2i}^e(x)]$ are the vectors of distributed source terms representing the effect of an external exciting electromagnetic field. These terms are equal to zero in the absence of an external exciting electromagnetic field.

In (20), there are neglected the terms corresponding to wire impedance and the ground admittance. Indeed, it has been shown in literature [30–33] that for a typical frequency range of interest (bellow 10 MHz) and for typical overhead lines, these parameters can be disregarded with reasonable approximation.

The expression for the mutual ground impedance between two conductors i and j has been derived by Sunde and is given by:

$$\underline{Z}_{gij} = \frac{j\omega\mu_0}{\pi} \cdot \int_0^\infty \frac{e^{-(h_i+h_j)\cdot x}}{\sqrt{x^2+\underline{\gamma}_g^2}+x} \cdot cos\,(r_{ij}x)dx \tag{21}$$

where:

$$\underline{\gamma}_g = \sqrt{j\omega\mu_0 \cdot (\sigma_g + j\omega\varepsilon_0\varepsilon_{rg})} \tag{22}$$

in which σ_g and ε_{rg} are the ground conductivity and relative permittivity.

The diagonal terms of the ground impedance matrix are given by:

$$\underline{Z}_{gij} = \frac{j\omega\mu_0}{\pi} \cdot \int_0^\infty \frac{e^{-2h_i \cdot x}}{\sqrt{x^2 + \underline{\gamma}^2} + x} \cdot dx \tag{23}$$

Eqs. (22) and (23) are not suitable for a numerical evaluation since they involve Sommerfeld integrals. Several approximate expressions for $\underline{Z}_{gij}$ have been presented in literature, but one of the simplest forms was proposed by Sunde himself and is given by the following logarithmic function [29]:

$$\underline{Z}_{gij} \cong \frac{j\omega\mu_0}{2\pi} \cdot ln \frac{1 + \underline{\gamma}_g \cdot h_i}{\underline{\gamma}_g \cdot h_i} \tag{24}$$

It has been shown [33] that (24) is an excellent approximation of the general expression (23).

In the following, the logarithmic approximation is extended also to off-diagonal terms. Using Euler relation in (21) and after some simple mathematical manipulations, we can obtain [33]:

$$\underline{Z}_{gij} = \frac{j\omega\mu_0}{\pi} \cdot \left(\frac{1}{2} \int_0^\infty \frac{e^{-2\underline{h}_{ij} \cdot x}}{\sqrt{x^2 + \underline{\gamma}_g^2} + x} \cdot dx + \frac{1}{2} \int_0^\infty \frac{e^{-2\underline{h}_{ij}^* \cdot x}}{\sqrt{x^2 + \underline{\gamma}_g^2} + x} \cdot dx \right) \tag{25}$$

in which $\underline{h}_{ij}$ is a complex quantity, and $\underline{h}_{ij}^*$ is its complex conjugate [31]:

$$\underline{h}_{ij} = \frac{h_i + h_j}{2} + j \cdot \frac{r_{ij}}{2} \tag{26}$$

Using the earlier mentioned approximate identity between expressions (23) and (24) we can express:

$$\begin{aligned} \frac{j\omega\mu_0}{2\pi} \cdot \int_0^\infty \frac{e^{-2\underline{h}_{ij} \cdot x}}{\sqrt{x^2 + \underline{\gamma}_g^2} + x} \cdot dx &\cong \frac{j\omega\mu_0}{4\pi} ln \left(\frac{1 + \underline{\gamma}_g \cdot \underline{h}_{ij}}{\underline{\gamma}_g \cdot \underline{h}_{ij}} \right) \\ \frac{j\omega\mu_0}{2\pi} \cdot \int_0^\infty \frac{e^{-2\underline{h}_{ij}^* \cdot x}}{\sqrt{x^2 + \underline{\gamma}_g^2} + x} \cdot dx &\cong \frac{j\omega\mu_0}{4\pi} ln \left(\frac{1 + \underline{\gamma}_g \cdot \underline{h}_{ij}^*}{\underline{\gamma}_g \cdot \underline{h}_{ij}^*} \right) \end{aligned} \tag{27}$$

Introducing (27) in (25), the following approximation can be derived for the general term $\underline{Z}_{gij}$ of the ground impedance matrix:

$$\underline{Z}_{gij} = \frac{j\omega\mu_0}{\pi} \cdot \left(\frac{1}{2} \int_0^\infty \frac{e^{-2\underline{h}_{ij} \cdot x}}{\sqrt{x^2 + \underline{\gamma}_g^2} + x} \cdot dx + \frac{1}{2} \int_0^\infty \frac{e^{-2\underline{h}_{ij}^* \cdot x}}{\sqrt{x^2 + \underline{\gamma}_g^2} + x} \cdot dx \right) \tag{28}$$

By introducing another type of approximation called the low-frequency approximation $(\sigma_g > > \omega\varepsilon_0\varepsilon_{rg})$, valid for low-frequency analysis, the general expression will become [33]:

$$\underline{Z}_{gij} = \frac{j\omega\mu_0}{\pi} \cdot \int_0^{\infty} \frac{e^{-(h_i+h_j)\cdot x}}{\sqrt{x^2 + j\omega\mu_0\sigma_g} + x} \cdot cos\,(r_{ij} \cdot x)dx \tag{29}$$

And in particular, the diagonal terms will be given by:

$$\underline{Z}_{gij} = \frac{j\omega\mu_0}{\pi} \cdot \int_0^{\infty} \frac{e^{-2h_i\cdot x}}{\sqrt{x^2 + j\omega\mu_0\sigma_g} + x} \cdot dx \tag{30}$$

Therefore, the general expressions for the elements of the ground impedance matrix in the multi-conductor transmission line equations are in terms of infinite integrals. Accurate approximations for the diagonal terms of the ground impedance matrix have already been presented in the literature [30–33] and that proposed by Sunde is the most accurate and simple.

3.2 Combined field and circuit method for evaluating inductive and capacitive matrices

In case of steady state or phase to ground fault HVPL operating conditions when the fault is far away of the common HVPL-MP right-of-way and a single frequency analysis is enough (no transients study is required), a simpler distributed elements equivalent electrical circuit approach could be implemented. At the same time for more complex conductor shapes (not only cylindrical/cable type conductors), a hybrid/combined field and circuit method should be applied [2, 3, 5, 25].

The hybrid method [2] was developed for single frequency analysis but could be extended to multiple frequency studies also if necessary. It combines the finite element method (FEM) analysis of the electromagnetic field in the common distribution corridor, with Faraday's law and with electric circuits theory respectively to determine the self and mutual inductances/capacitances between any metallic structures present in a HVPL-MP interference problem [5].

This type of analysis consists of solving Maxwell's equations in nonuniform three-dimensional space and consists of the following steps [5, 25]:

Step 1. A multi-conductor system that includes the pipeline, phase wires, and overhead ground wires together with electrical towers and grounding systems is modeled.

Step 2. A reference current/voltage is set on phase wire A $(0°)$ and zero value currents/voltages on the rest of the metallic conductors (including the other phases).

Step 3. The magnetic vector potential $\vec{A}$ on the cross sections of all the metallic structures ($\vec{A}_{cond}$) that make up the studied problem is determined by FEM analysis.

Step 4. The electric charges Q, due to the imposed reference voltage, that appear on the surface of all the conductors, respectively on the surface of the ground, are determined by means of the FEM analysis.

Step 5. Self-inductance per unit length of the conductor on which the reference current I_{ref} was imposed is determined, respectively, the mutual inductances per unit length given by this current in the rest of the metallic structures are computed.

$$L_{self} \text{ or } L_{mut} = \frac{\vec{A}_{cond}}{\vec{I}_{ref}}\,[\mathrm{H/m}] \tag{31}$$

Step 6. Based on the determined electrical charges Q or linear charge distributions ρ_ℓ, the conductor-soil (self) capacity per unit length of the conductor on which the reference voltage V_{ref} was imposed, respectively, the mutual capacities compared to the rest of the conductors per unit length, are evaluated

$$C_{self} \text{ or } C_{mut} = \frac{\rho_\ell}{U} = \frac{\rho_\ell}{V_{ref} - V_0} = \frac{\rho_\ell}{V_{ref}} [\text{C/m}] \tag{32}$$

Step 7. Steps 3 and 5 respectively 4 and 6 are repeated for the case where the imposed reference current/voltage is set one by one on all the metallic structures that constitutes the investigated interference problem.

Using this iterative algorithm [25], the matrixes of self and mutual inductances and capacities between all the present structures could be constructed. Thus, the obtained inductive and capacitive matrixes describe the inductive respectively the capacitive couplings mechanism between HVPL and MP. Therefore, they could be used to solve the equivalent electric circuit model [5, 25].

The FEM evaluates the total electromagnetic interference effect in a single step, avoiding the separation of the inductive, capacitive, and conductive components, which is necessary in the circuit model. However, the FEM limitation is that when the common corridor is very long and consists of many circuits, the modeling and computation can be extremely time-expansive.

Therefore, usually 2D cross-section FEM analysis is applied for HVPL-MP parallel exposures. The following system of equations describes the linear 2D electromagnetic diffusion problem for the z-direction (along the common right-of-way) components A_z of the magnetic vector potential and J_z of the total current density vector [2]:

$$\begin{cases} \dfrac{1}{\mu_0 \mu_r} \cdot \left[\dfrac{\partial^2 A_z}{\partial x^2} + \dfrac{\partial^2 A_z}{\partial y^2} \right] - j\omega\sigma A_z + J_{sz} = 0 \\ -j\omega\sigma A_z + J_{sz} = J_z \\ \iint_{S_i} J_z ds = I_i \end{cases} \tag{33}$$

where J_{sz} is the source current density in the z direction, and I_i is the imposed current on conductor i of S_i cross section.

Eq. (33) could be solved by any dedicated finite element calculation software to compute the magnetic vector potentials on the surface of each metallic structure (phase wires, sky wires and pipeline).

4. Analysis of main parameters that influence the induced voltages

For the numerical evaluation of the induced voltages in metallic pipelines exposed to the electromagnetic fields produced by nearby high-voltage power lines, several software applications and software packages that implement the above presented approaches or similar methodologies could be used. There are to main types of such software applications:

- Electromagnetic Transients Programs such as EMTP-RV, ATP-EMTP, or PSCAD, which are dedicated to steady-state and transient-state analysis of complex power networks and systems. These applications require the user to build up the circuit model of the analyzed system and to select the proper circuit elements and equipment.

- Partial Element Equivalent Circuit (PEEC) technique-based software packages such as CDGES or XGSLab that are dedicated for the investigation of multi-conductor or grounding grid system. These software packages usually require the user to draw the analyzed multi-conductor system and to set the corresponding material parameters.

To highlight the effect of the main parameters that could influence the level of induced voltages in metallic pipelines, a usual HVPL-MP electromagnetic interference problem is simulated and analyzed in the following using the EMTP-RV software, due to its user-friendly interface and high computational capabilities.

An underground metallic gas transportation pipeline is considered to share the same distribution corridor with a single circuit 220 kV/50 Hz overhead electrical power line.

To model the electromagnetic coupling between HVPL and the underground MP, the new *Line/Cable Data* component is used from EMTP-RV, which allows to combine overhead conductors (HVPL phase wires and sky wires) with above or underground cables (the pipeline is considered as an unenergized insulated hollow cable with the same geometrical dimensions).

Using this *Line/Cable Data* component, the equivalent *wideband* transmission line model of the common distribution corridor is created (see **Figure 11**). The *wideband* transmission line model is based on the Universal Line Model (ULM) introduced in [34] applying the transmission line approach described in section 3.1. The *wideband* model

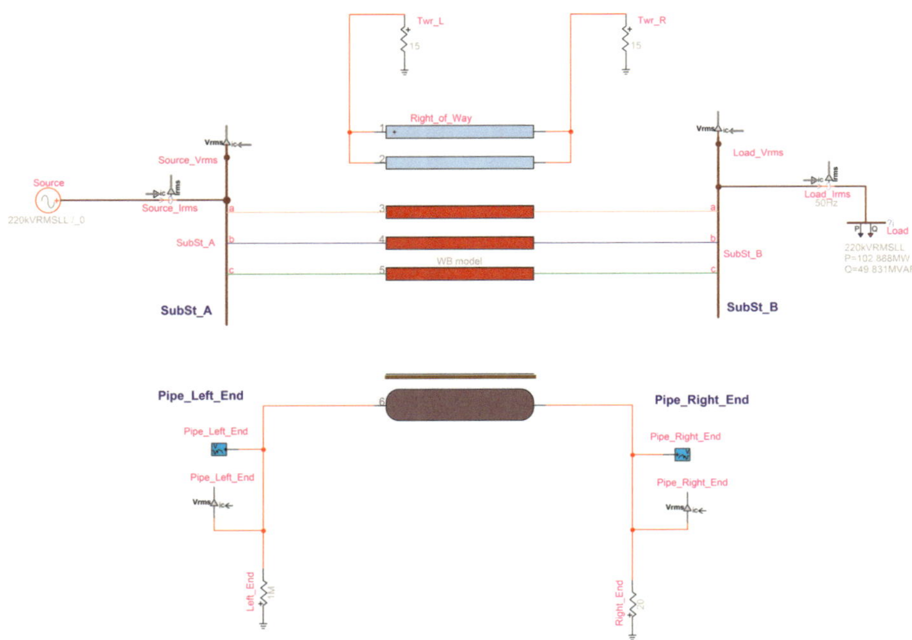

Figure 11.
General overview of the HVPL-MP common distribution corridor implemented in EMTP-RV.

form EMTP-RV uses complex poles and zeros for the rational approximation of the frequency-dependent characteristic admittance and wave propagation matrices, while it also takes fully into account the frequency dependence of the modal transformation matrix [23]. Therefore, it can be considered the most accurate time-domain model, its applicability being extended to any type of overhead or underground transmission line.

The HVPL consists of three phase wires in a horizontal layout with a 7.4 m spacing between each other placed at 24 m above ground and two sky wires placed at 26.5 m above ground with a 4.4 m spacing from HVPL axis. The metallic pipeline is buried at 1.2 m depth. The geometrical and electrical parameters of the conductors involved in the analyzed HVPL-MP common distribution corridor (phase wires, sky wires, and the metallic pipeline) are presented in **Table 1**.

Figure 11 presents the electrical circuit model implemented in the EMTP-RV software of the investigated HVPL-MP electromagnetic interference problem considering a single equivalent transmission line element for the entire common distribution corridor. Such an implementation allows us to evaluate only the maximum values of the induced voltage, which is reached at the ends of the analyzed pipeline.

To be able to evaluate the variation of the induce voltages along the pipeline length and not just the maximum values, the common distribution corridor must be divided into several consecutive transmission line sections (segments). Usually, the length of such a transmission line segment is considered equal the span between two consecutive power line towers. This approach allows also to investigate the influence of power line towers and their grounding in case of transient state faults like the ones generated by lightning strikes to HVPL. **Figure 12** presents a detailed representation of a 1 km long HVPL-MP common right-of-way using 250 m transmission line segments as the distance between two consecutive power line towers.

To be as close as possible to real-life situations, at both ends of common distribution corridor, the pipeline is continued with 5 km long transmission line segments representing the pipeline outside of the HVPL zone of influence. Being an underground pipeline, the inductive coupling will prevail in the electromagnetic interference problem that occurs.

4.1 Steady-State operating condition

As a first step, the induced AC voltages are evaluated in the underground MP during power line steady-state conditions with a 300 A symmetrical current load (around 115 MVA power load with a 0.9 power factor). The HVPL-MP distribution

Source phase voltage (kV)	**220**
Internal radius of the pipeline (cm)	38.1
External radius of the pipeline (cm)	39.1
Thickness of the pipeline's coating (cm)	1
Resistivity of the pipeline's metal (Ohm/m)	1.72E-7
Relative permeability of the pipeline's metal	300
Relative permittivity of the pipeline's coating	2.3

Table 1.
Parameters that describe the system under investigation.

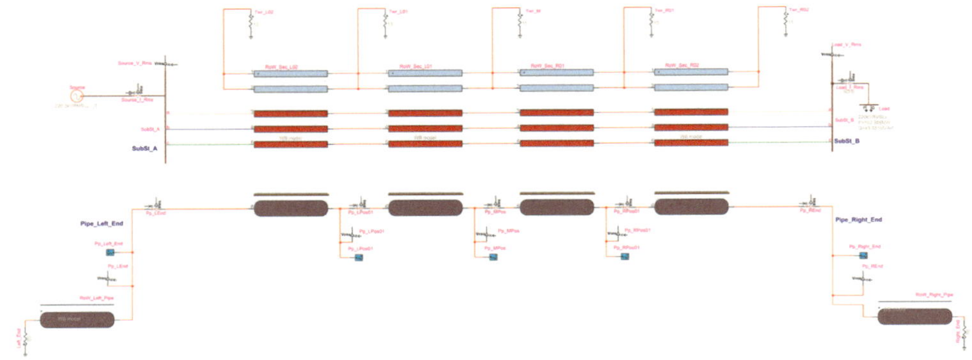

Figure 12.
Detailed representation of a HVPL-MP common distribution corridor implemented in EMTP-RV.

corridor has a length of 2 km. The underground MP follows a parallel route within the right-of-way at 30 m separation distance from the power line axis on the right side. The HVPL phase wires are placed in the phase A (0°), phase B (−120°), and phase C (−240°) order on the power line towers. Therefore, phase C (−240°) is the nearest phase wire to the underground MP. The soil is considered uniform with a resistivity of 100 Ωm for the entire right-of-way.

The time-domain variation of the induced voltages at different locations along pipeline length in the investigated common distribution corridor evaluated through EMTP-RV simulation and modeling is showcased in **Figure 13**.

The graphical representation from **Figure 13** highlights values around 6.5 V regarding the induced AC voltages, which are in dangerous range from corrosion

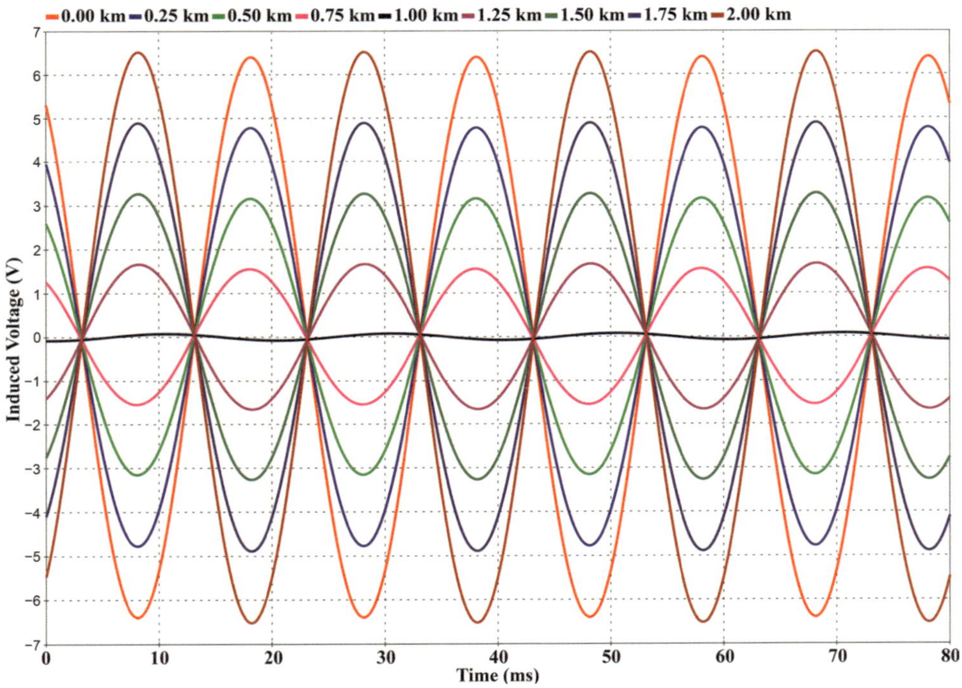

Figure 13.
Time-domain representation of the AC induced voltages in MP, for steady-state HVPL condition.

point of view. The critical values of the induced voltages, regarding the start of electrochemical corrosion reaction, are around 1.2 V as in [9, 10, 22] are presented.

Due to the induced voltages in the metallic pipeline, leakage currents will appear that depend directly on the induced voltage levels, the pipeline conductivity, and its insulation. These leakage currents could highly increase if pipeline insulation defects are present, accelerating the corrosion process.

Figure 14 shows the variation of the induced AV voltages along the pipeline length as RMS values. It can be observed that the "V"-shaped curve as reported in several literature studies [3, 17, 23–26], with the highest induced RMS voltage values recorded at right-of-way ends. It must be mentioned that at each time moment, the pipeline has opposite electrical potential at its end as in **Figure 13** can be noticed.

4.2 Parametric analysis

To highlight the effect that different geometric and electrical parameters can have on the values of the induced voltages in the case of a HVPL-MP interference problem, a parametric analysis is made, by taking into account: the length of the common distribution corridor, the HVPL-MP separation distance, the HVPL load current, and the soil resistivity variation.

Figure 15 highlights the induced voltages along the MP in the case of a right-of-way with a length that varies from 0.5 km to 10 km. It is found that the induced voltage increases quasi-proportionally with the length of the common distribution corridor, up to a "critical" (or characteristic) length that depends on the geometrical and electrical parameters of the pipeline, the insulation, and the surrounding environment (soil), being defined based on the propagation constant of the electromagnetic wave along the pipeline. Among these parameters, various studies have shown that soil resistivity has the least influence.

Another important aspect in an HVPL-MP interference problem is the distance between the electrical line and the metallic structure. **Figure 16**.a highlights the induced voltage values in the MP if it is positioned at up to 1000 meters from the power line (on either side of it). While **Figure 16**.b is a more detailed representation of the induced voltages in MP recorded near the metallic towers of the power line, up to 100 m on either side HVPL axis.

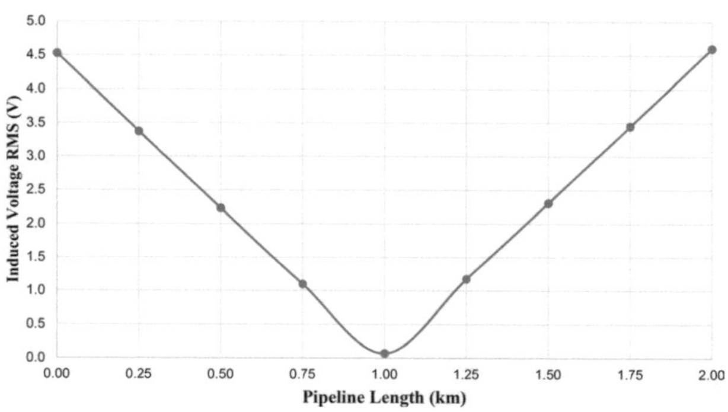

Figure 14.
Induced voltages along the MP, in case of steady-state condition.

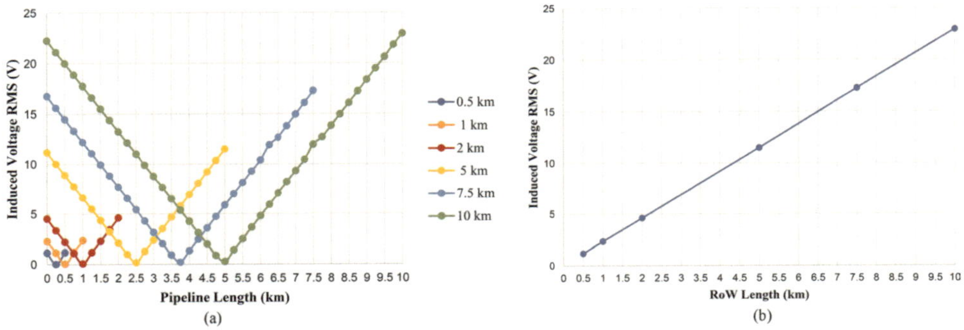

Figure 15.
(a) Induced voltages along pipeline length for different right-of-way length, and (b) Maximum induced voltage variation with right-of-way length.

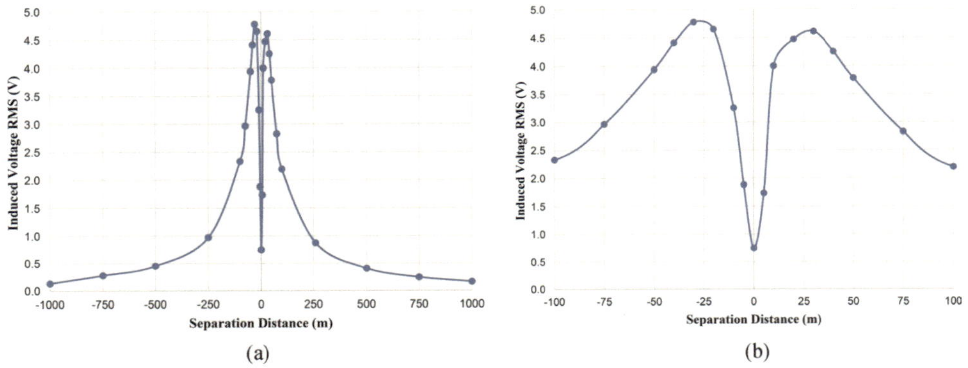

Figure 16.
(a) Maximum induced voltage variation with HVPL-MP separation distance, and (b) Highlight for close HVPL-MP exposures (under 100 m).

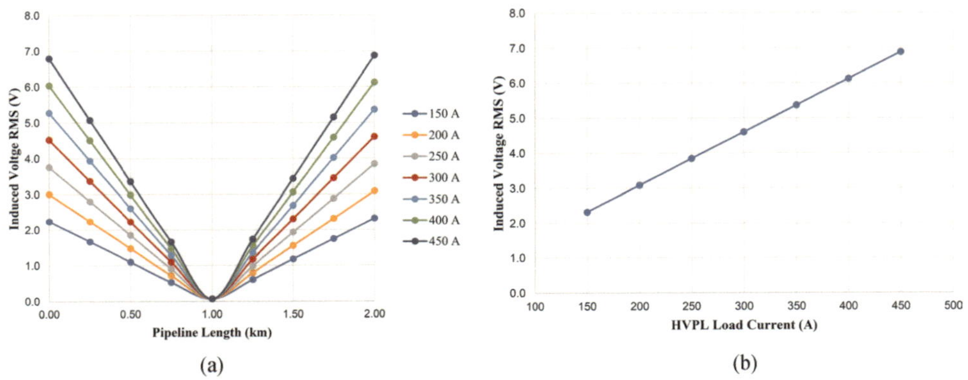

Figure 17.
(a) Induced voltages along pipeline length for different HVPL load currents, and (b) maximum induced voltage variation with HVPL load current.

To investigate the induced voltage levels that can occur for different power flows, the symmetrical current load on HVPL was considered in the range of 150–450 A. According to **Figure 17**, the levels of the induced voltages in MP increase with the value of current load (an increase from 300 A to 450 A of the load current leads to 50% increase of the maximum value of the induced AC voltage).

Evaluation of induced voltages for different soil resistivity values has shown than even when the soil resistivity varies in a large range like from 10 Ωm to 10 kΩm, the maximum induced voltage values vary only with 2–3% with regard to the steady-state results. The soil resistivity has a higher influence on the pipeline leakage currents in case of insulation defects or in case of lighting generated HVPL faults when soil ionization phenomena could occur around power line tower groundings.

4.3 Oblique exposures and intersections HVPL-MP

The above-presented EMTP-RV implementation and the parametric analysis consider perfect parallel exposure between HVPL and MP. However, in practice [20, 35, 36], there are frequent situations in which above or underground MPs present oblique exposures to the HVPL axis (see **Figures 18** and **19**) [8].

Therefore, to analyze real-life HVPL-MP interference problems using electromagnetic transients programs like EMTP-RV, each oblique HVPL-MP exposure section has to be replaced with an equivalent parallel HVPL-MP exposure section for which the HVPL-MP separation distance d_{eq} is given by the Eq. (34) as long as the ratio between d_2 and d_1 (see **Figure 18**) is less than or equal to 3 [8].

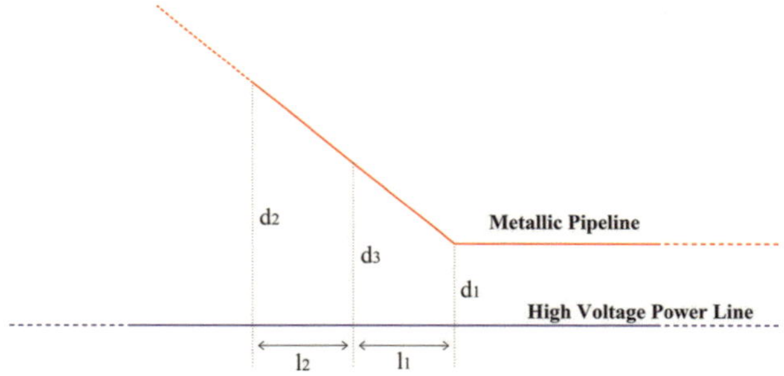

Figure 18.
Oblique HVPL-MP exposure.

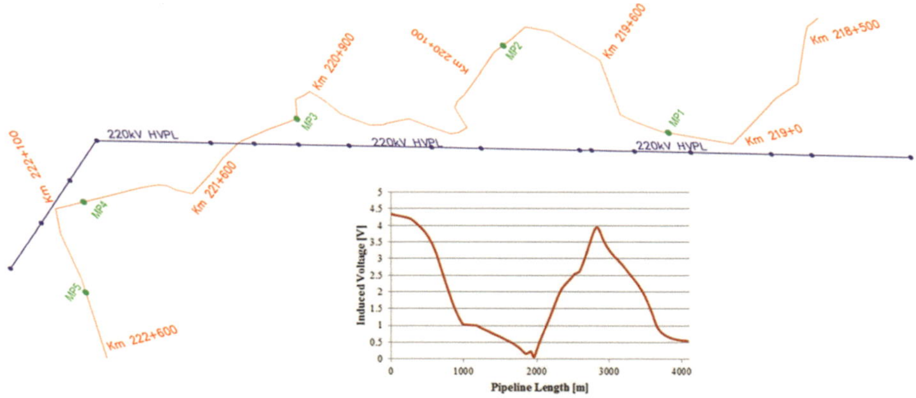

Figure 19.
Induced AC voltage in case of a real HVPL-MP right-of way [37].

$$d_{eq} = \sqrt{d_1 \cdot d_2} \tag{34}$$

If the ratio between the distances at which the two ends of the oblique exposure d_2/d_1 is greater than 3, the oblique exposure must be divided into two subsections, both sections verifying the imposed condition:

$$\frac{d_3}{d_1} \leq 3 \quad \text{and} \quad \frac{d_2}{d_3} \leq 3 \tag{35}$$

On the other hand, if MP sub-crosses the HVPL, this section is replaced by an equivalent section, in which the MP is located at $d = 6\ m$ distance from the HVPL, for a length equal to the length of the MP projection segment considered in around the crossing point up to a separation distance less than or equal to 10 m on either sides of the HVPL.

Figure 19 presents the AC induce voltage variation along a complex real-life HVPL-MP common distribution corridor applying the abovementioned right-of-way segmentation procedure [37].

5. Conclusions

An evaluation of the electromagnetic interferences phenomena effects that affect metallic pipelines placed in the vicinity of high-voltage power lines is presented in this chapter highlighting different analyzing and modeling techniques.

As it is shown, the presence of AC power supply systems may cause voltages to build up in nearby metallic pipeline systems, due to one or more of the following mechanisms: inductive, conductive and capacitive coupling. Such voltages may put the pipeline operating personal in danger, damage the pipeline, disturb the electrical/ electronic equipment connected to the pipeline.

The induced voltages and currents on the buried pipelines may be dangerous to pipeline security due to the AC corrosion and deteriorated the cathodic protection devices. Consequently, a mitigation system is required to be designed to reduce the effects of the corrosion of which main causes are the leakage current densities from the pipeline into the surrounding soil due to AC-induced voltages.

Since even very high-quality insulation of underground metal pipelines could present insulation defects, to reduce the probability of occurrence of corrosion phenomena, European standards [14] suggest limiting the induced voltages depending on the electrochemical characteristics of the soil, so that they do not exceed:

- 10 V if the soil resistivity is greater than 25 Ωm;
- 4 V if the soil resistivity is lower than 25 Ωm.

Different techniques can be applied to mitigate induced AC voltages and gas, such as cancelation wires, gradient control wires, and insulating joints [9, 10, 26].

Thanks

Energy Transition Research Center (EnTReC) of Technical University of Cluj-Napoca is an EMTP-RV official educational partner and would like to thank POWERSYS SOLUTIONS for offering the academic license of the software.

Author details

Denisa Șteț*, Levente Czumbil and Dan Doru Micu
Energy Transition Research Center (EnTReC), Technical University of Cluj-Napoca (TUCN), Cluj-Napoca, Romania

*Address all correspondence to: denisa.stet@ethm.utcluj.ro

References

[1] Dawalibi FP, Southey RD. Analysis of electrical interference from power lines to gas pipelines. I. Computation methods. IEEE Transaction on Power Delivery. 1989;**4**(3):1840-1846. DOI: 10.1109/61.32680

[2] Christoforidis GC, Labridis DP, Dokopoulos PS. A hybrid method for calculating the inductive interference caused by faulted power lines to nearby buried pipelines. IEEE Transaction on Power Delivery. 2005;**20**(2):1465-1473. DOI: 10.1109/TPWRD.2004.839186

[3] Micu DD, Christoforidis GC, Czumbil L. AC interference on pipelines due to double circuit power lines: A detailed study. Electric Power Systems Research. 2013;**103**:1-8. DOI: 10.1016/j.epsr.2013.04.008

[4] Charalambous CA, Demetriou A, Lazari A. Effects of electromagnetic interference on underground pipelines caused by the operation of high voltage AC traction systems: The impact of harmonics. IEEE Transaction on Power Delivery. 2018;**33**(6):2664-2672. DOI: 10.1109/TPWRD.2018.2803080

[5] Șteț D. Contributions to Methods of Modeling and Prediction of AC Electromagnetic Interferences. Cluj-Napoca: Technical University of Cluj-Napoca; 2010

[6] Goidanich S, Lazzari L, Ormellese M. AC corrosion. Part 1: Effects on overpotentials of anodic and cathodic processes. Corrosion Science. 2010; **52**(2):491-497. DOI: 10.1016/j.corsci.2009.10.005

[7] Goidanich S, Lazzari L, Ormellese M. AC corrosion. Part 2: Parameters influencing corrosion rate. Corrosion Science. 2010;**52**(3):916-922. DOI: 10.1016/j.corsci.2009.11.012

[8] CIGRE. Guide on the Influence of the High Voltage AC Power System on Metallic Pipelines. Paris, France: CIGRE, WG 36.02; 1995

[9] CEOCOR. AC Corrosion on Cathodically Protected Pipelines—Guidelines for Risk Assessment and Mitigation Measures. Luxembourg: CEOCOR. 2021. Available from: https://ceocor.lu/download/AC-Corrosion-Booklet-on-cathodically-protected-pipelines-ed-2001.pdf

[10] EN ISO 18086:2021—Corrosion of Metals and Alloys—Determination of AC Corrosion—Protection Criteria

[11] EN 50443:2012—Effects of Electromagnetic Interference on Pipelines caused by High Voltage A.C. Electric Traction systems and/or High Voltage A.C. Power Supply Systems

[12] Czumbil L. Software Package Developments Based on Modern Analysis Techniques of the Electromagnetic Interference Problems between Electrical Power Lines and Nearby Metallic Pipelines. Cluj-Napoca: Technical University of Cluj-Napoca; 2012

[13] Henry WO. Electromagnetic Compatibility Engineering. Chichester: Wiley; 2009. p. 880. DOI: 10.1002/9780470508510

[14] Coelho L, Sotelo G, Lima ACS. Detailed versus simplified representation of a pipeline for assessment of inductive and conductive couplings to an overhead transmission lines during steady-state and fault conditions. International Journal of Electrical Power & Energy Systems. 2022;**142**(B):108350. DOI: 10.1016/j.ijepes.2022.108350

[15] Gupta A, Thomas MJ. Coupling of high voltage AC power line fields to metallic pipelines. In: Proceedings of the 9th International Conference on Electromagnetic Interference and Compatibility (INCEMIC 2006). New York: IEEE; 2010

[16] Christoforidis GC, Labridis D, Dokopoulos P. Inductive interference calculation on imperfect coated pipelines due to nearby faulted parallel transmission lines. Electric Power Systems Research. 2003;**66**(2):139-148. DOI: 10.1016/S0378-7796(03)00018-X

[17] Lucca G. Different approaches in calculating AC inductive interference from power lines on pipelines. IET Science, Measurement & Technology. 2018;**12**(6):802-806. DOI: 10.1049/iet-smt.2018.0086

[18] Wang Y, Wang X, Zhu Z, Zeng L, Yong J. Effects of harmonic induction on metallic pipeline caused by overhead power lines. International Journal of Electrical Power & Energy Systems. 2022;**137**:107758. DOI: 10.1016/j.ijepes.2021.107758

[19] Şteţ D, Czumbil L, Micu DD, Ţopa V, Ancăş L. Stream gas pipeline in proximity of high voltage power lines. Part I—Soil resistivity evaluation. In: Proceedings of the 47th International Universities' Power Engineering Conference (UPEC 2012); 04-07 September 2012. New York: IEEE; 2012. DOI: 10.1109/UPEC.2012.6398445

[20] Czumbil L, Şteţ D, Micu DD, Ţopa V, Ancăş L. Stream gas pipeline in proximity of high voltage power lines. Part II—Induced voltage evaluation. In: Proceedings of the 47th International Universities' Power Engineering Conference (UPEC 2012); 04-07 September 2012. New York: IEEE; 2012. DOI: 10.1109/UPEC.2012.6398444

[21] Haubrich HJ, Flechner BA, Machczynski W. A universal model for the computation of the electromagnetic interference on earth return circuits. IEEE Transaction on Power Delivery. 1994;**9**(3):1593-1599. DOI: 10.1109/61.311205

[22] Li L, Gao X. AC corrosion interference of buried long distance pipeline. In: Proceedings of the 3rd International Conference on Intelligent Control, Measurement and Signal Processing and Intelligent Oil Field (ICMSP 2021); 23–25 July 2021. New York: IEEE; 2021. DOI: 10.1109/ICMSP53480.2021.9513377

[23] Papadopoulos TA, Chrysochos AI, Doukas DI, Papagiannis GK, Labridis DP. Induced voltages and currents: Overview and evaluation of simulation models and methodologies. In: Proceedings of Mediterranean Conference on Power Generation, Transmission, Distribution and Energy Conversion (MedPower 2016). Belgrade; 2017

[24] Christoforidis G, Labridis D. Inductive interference on pipelines buried in multilayer soil due to magnetic fields from nearby faulted power lines. IEEE Transaction on Electromagnetic Compatibility. 2005;**47**(2):254-262. DOI: 10.1109/TEMC.2005.847399

[25] Czumbil L, Christoforidis GC, Micu DD, Şteţ D, Ceclan A, Pop O. A user-friendly software application for induced A.C. interference evaluation. In: Proceedings of the 46th International Universities' Power Engineering Conference (UPEC 2011); 05-08 September 2011. New York: IEEE; 2012

[26] Cristofolini A, Popoli A, Sandrolini L. Numerical modelling of interference from AC power lines on buried metallic pipelines in presence of

mitigation wires. In: Proceedings of the IEEE International Conference on Environment and Electrical Engineering and IEEE Industrial and Commercial Power Systems Europe (EEEIC/I&CPS Europe 2018); 12–15 June 2018. New York: IEEE; 2018

[27] Nakagawa M, Ametani A, Iwamoto K. Further studies on wave propagation in overhead lines with earth return: Impedance of stratified earth. Proceedings of the Institution of Electrical Engineers. 1973;**120**(12): 1521-1528. DOI: 10.1049/piee.1973.0312

[28] Popoli A, Sandrolini L, Cristofolini A. Inductive coupling on metallic pipelines: Effects of a nonuniform soil resistivity along a pipeline-power line corridor. Electrical Power Systems Research. 2020;**189**: 106621. DOI: 10.1016/j.epsr.2020.106621

[29] Sunde ED. Earth Conduction Effects in Transmission Systems. New York: Van Nostrad Company; 1949. p. 373

[30] Velasquez R, Reynolds PH, Mukhedkar D. Earth-return mutual coupling effects in ground resistance measurement of extended grids. IEEE Transactions on Power Apparatus and Systems. 1983;**102**(6):1850-1857

[31] Sarmiento HG, Mukhedkar D, Ramanchandran V. An extension to the study of earth return mutual coupling effects in ground impedance field measurement. IEEE Transactions on Power Delivery. 1988;**3**(1):96-101. DOI: 10.1109/61.4232

[32] Rogers E, White JF. Mutual coupling between horizontal earth return conductors using actual routing parameters. IEEE Transactions on Power Delivery. 1990;**5**(3):1266-1274. DOI: 10.1109/61.57965

[33] Micu DD, Ceclan A, Dărăbant L, Şteţ D. Analytical and numerical development of the electromagnetic interference between a high-voltage power line and a metallic underground pipeline. In: Proceedings of the 8th International Symposium on Advanced Electromechanical Motion Systems & Electric Drives Joint Symposium (ELECTROMOTION 2009); 01-03 July 2009. New York: IEEE; 2009. DOI: 10.1109/ELECTROMOTION.2009.5259090

[34] Morched A, Gustavsen B, Tartibi M. A universal model for accurate calculation of electromagnetic transients on overhead lines and underground cables. IEEE Transactions on Power Delivery. 1999;**14**(3):1032-1038. DOI: 10.1109/61.772350

[35] Şteţ D, Micu DD, Czumbil L, Manea B. Case studies on electromagnetic interference between HVPL and buried pipelines. In: Proceedings of the International Conference and Exposition on Electrical and Power Engineering (EPE 2014); 16-18 October 2014. New York: IEEE; 2014. DOI: 10.1109/ICEPE.2014.6969903

[36] Mureşan A, Papadopoulos TA, Czumbil L, Chrysochos AI, Farkas T, Chioran D. Numerical modeling assessment of electromagnetic interference between power lines and metallic pipelines: A case study. In: Proceedings of the 9th International Conference on Modern Power Systems (MPS 2021); 16-17 June 2021. New York: IEEE; 2021. DOI: 10.1109/MPS52805.2021.9492630

[37] Şteţ D, Czumbil L, Erchedi M, Hanc S, Ancăş L. Electromagnetic interference issues in case of metallic structures. Acta Electrotehnica. 2013; **54**(5):470-473

Chapter 4

Assessment of the Structural Integrity of the Pipes with Anomalies Such as Local Elastic-plastic Deformations

Alin Dinita

Abstract

Pipelines are one of the most practical and economically efficient ways to transport dangerous and/or flammable substances, for which road or rail transport is often impossible. The evaluation of the processes that can negatively influence the performance of the pipelines is particularly important for assessing the risk associated with the operation of these technical systems and the potential for technical accidents. The anomalies that can be found on the pipes can be classified into two main categories. Imperfections that do not inadmissibly affect their load-bearing capacity and defects with significant negative influences on the correct operation and load-bearing capacity of the piping, which require supervision and maintenance measures. The influence of these anomalies and the processes that lead to the decrease of the pipeline-bearing capacity constitutes the main objectives of the analysis performed. The local elastic-plastic deformation anomalies are considered, for which the parameters of the geometric model of the defect profile, the conditions for generating these anomalies, and the evaluation methods, respectively, were analyzed, both analytically and experimentally.

Keywords: pipeline, defect assessment, elasto-plastic deformations, burst pressure, dent and gouge

1. Introduction

The probability of the appearance of anomalies on the pipelines is closely related to the mechanical resistance and tenacity characteristics of the material (steel) from which the pipelines are made. Anomalies that can be found on pipelines can be classified into material loss anomalies, which consist of the thinning of the pipeline wall through the loss of metal in the presence or absence of a corrosive process, and anomalies such as cracks and anomalies such as local elastic-plastic deformations.

The steels from which the pipelines are mainly made have the ferrito-pearlitic structure typical of carbon or low-alloy hypoeutectoid steels, and the increase in their mechanical resistance characteristics is achieved mainly by increasing the carbon concentration, which has the effect of increasing the percentage content of perlite in the structure. Modern steels for pipes have a structure with acicular ferrite (low carbon bainite), their high mechanical resistance characteristics being achieved mainly by obtaining a very fine grain and ensuring hardening effects by the precipitation of some intermetallic compounds. The weldability of these steels is satisfactory if the carbon concentration is not increased excessively and if the welding procedure and regime are chosen appropriately.

The carbon concentration of these steels does not exceed 0.30...0.31%, to obtain the higher degrees of resistance resorting to the use of manufacturing recipes with manganese concentrations higher than those typical of carbon steels. Flat strip semi-finished products intended for the manufacture of longitudinally or helically welded pipes are made from such steels by controlled rolling or thermomechanical rolling (rolling with high degrees of deformation in which working temperatures, heating, cooling, and deformation speeds are strictly controlled), which emphasize the presence in these steels of a wide range of microalloying elements (Nb, V, Ti, Mo, etc.).

Defining and classifying the factors that lead to the degradation and failure of pipelines is a problem for which specialized literature provides a multitude of solutions. To carry out the analysis of the factors and processes that lead to the progressive degradation and failure of pipelines, it was considered that the factors to be considered correspond to the potential hazards specified by the ASME B31.8S standard for risk assessment and development of pipeline integrity management plans intended for natural gas transport.

The classification of these factors (potential hazards) is presented synthetically in **Table 1**; as can be seen, the ASME B31.8S standard recommends the division into three classes of factors that can determine the failure of pipelines, and for each class, there are three categories of factors, each category having one or more factors.

Imperfections are anomalies in configuration, dimensions, microstructure, etc. present in the pipelines, which do not inadmissibly affect their load-bearing capacity, and the defects are the imperfections with significant negative influences on the correct operation and the load-bearing capacity of the pipelines, which require maintenance measures. The classification of pipe imperfections and defects can be based on the configuration criterion that defines the following four categories of pipe imperfections and defects:

1.1 Imperfections and geometric defects produced by the local deformation of the pipes

Local deformations or dents are deviations from the circular shape of the cross-section of the piping, obtained by local deformation of the piping, inwards, without removing the material and, as a result, without reducing the wall thickness. Indentations influence the flow of gases in the pipelines and can cause major difficulties in performing cleaning or washing operations and in the internal inspection of the pipelines, by blocking the movement of working tools or PIG devices. Indentations of an elastic nature, produced, for example, by the interaction of pipes with pieces of rock can be eliminated by simply removing the cause (pieces of rock that produced the deformation); *Gouges* are areas of the pipes where the wall thickness has been locally reduced due to the removal of material through a mechanical action, and gouges are defects or imperfections of great severity because their presence can lead to the initiation of brittle cracking processes, representative types of such defects are shown in **Figure 1**.

Category	Type	Factor name	Category	Type	Factor name
A. Time-dependent			*B. Stable*		
1	a	external corrosion	*manufacturing related defects*		
2	a	internal corrosion	1	a	defective pipe seam
3	a	stress corrosion cracking	1	b	defective pipe
			welding/fabrication related		
C. Time-independent			2	a	defective pipe girth weld
third party / mechanical damage			2	b	defective fabrication weld
1	a	damage inflicted by first, second, or third parties	2	c	wrinkle bend or buckle
1	b	previously damaged pipe	2	d	stripped threads / broken pipe
1	c	vandalism	*equipment*		
2	a	incorrect operational procedure	3	a	gasket O-ring failure
weather-related and outside force			3	b	control equipment malfunction
3	a	cold weather	3	c	seal / pump packing failure
3	b	lightning	3	d	miscellaneous
3	c	heavy rains			
3	d	earth movements			

Table 1.
The categories of factors that can affect the integrity of natural gas pipelines.

Figure 1.
Imperfections and defects of local elastic-plastic deformation type.

1.2 Imperfections and defects such as material loss

These imperfections or defects consist in the general or local thinning of the wall of the pipeline through the loss of metal in the presence or absence of a corrosive process. The most common imperfections or defects in this category are as follows: *areas of local thinning* (areas on the surface of a pipe element, having the axial extension or length of the same order of magnitude as the circumferential extension or width, in which the material has been removed by corrosion and/or erosion), *pinching or pitting* (traces of local corrosion on the surface of a pipe element, in the form of cavities or holes, having the surface diameter of the ordinal size of the wall thickness of the respective pipe element), and representative types of such defects are shown in **Figure 2**.

1.3 Cracks

Cracks are the anomalies with the greatest harm, which produce strong mechanical stress concentration effects and significantly reduce the carrying capacity of the pipes.

Figure 2.
Imperfections and defects such as material loss.

DOI: http://dx.doi.org/10.5772/intechopen.108437

Their dimensions can change over time through stable growth, until they reach critical dimensions, at which unstable propagation and rupture of the tubing can occur. Depending on the toughness characteristics of the piping material, cracks can generate brittle fracture phenomena (which occur at high speeds and propagate over large distances, giving rise to damage with important consequences) or ductile fracture phenomena (which occur at small speeds and are preceded by plastic deformation processes, which consume an important part of the available energy and thus contribute to stopping the phenomenon and limiting its consequences), and representative types of such anomalies are presented in **Figure 3**.

Figure 3.
Cracks.

1.4 Other types of defects or imperfections

This category includes imperfections or defects that cannot be attached to one of the previously specified categories, for example: defects in the sealing systems of valves or fittings mounted on pipes, defects in threads, defects in the manufacture (lamination, welding) of pipes.

2. Characterization of the geometry of anomalies produced by local elastic-plastic deformation of pipes

The way of generating and modeling the geometry of the anomalies (imperfections and defects) produced by the local deformation of the pipes is a very delicate problem, considering both the complexity of the phenomena that take place in the deformed area and the need to obtain some methods for evaluating the bearing capacity of the pipes that present such anomalies.

In the specialized literature, you can find various geometric modeling (semi-elliptical, hemispherical, or hexagonal cavities) that take into account both the characteristics of the pipeline and the cause of these anomalies (third-party interventions). **Figure 4** shows two of the most important geometric models that characterize the imperfections and defects produced by local plastic deformation of pipes [1–3].

Local plastic deformation represents a change in the circular section of a pipe or a distortion of the pipe in a circular section. In **Figure 5**, a geometric modeling model of the shape of these elastic-plastic deformation type anomalies is proposed. The depth of this type of anomaly is defined as the maximum reduction of the diameter in the defect area compared to the initial nominal diameter of the pipe. This definition of deformation depth includes both local deformation and any distortion in the circular section of the pipe (ovality).

To determine the stress state in the area of an imperfection and defects produced by the local deformation of a pipe, it is necessary to carry out an analytical description of the geometry of the anomaly. For the indentations with axial orientation on the pipelines, the analytical description of the shape was proposed by using a cylindrical coordinate system, of the type defined in the sketch in **Figure 5** and a function with the following analytical form [4, 5]:

$$r(\varphi, z) = R - d_p(z)\left[1 + cos\left(2\pi\frac{\varphi}{\varphi_0(z)}\right)\right] |\varphi| \leq \varphi_0 |\varphi| > \varphi_0 \tag{1}$$

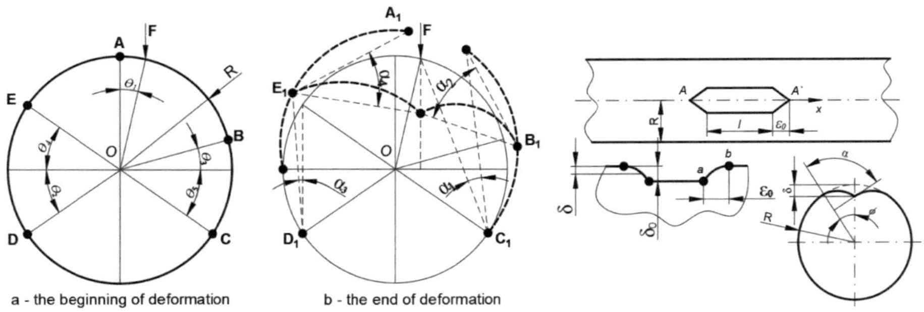

Figure 4.
Two models for defining the geometric profile for an anomaly of the type of local elastic-plastic deformations.

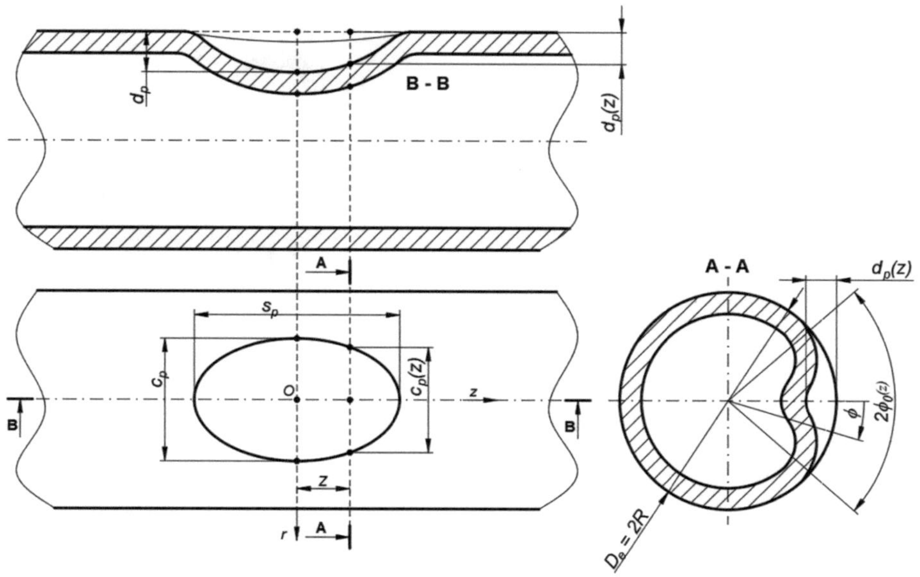

Figure 5.
Scheme of the description of the configuration of a pipeline in the area of a local elastic-plastic deformation type anomaly.

In Eq. (1) the circumferential anomalies profile are given as a deformation from the pipe radius R, the maximum of the deformation is given by the term $d_p(z)$, and the circumferential deformation of the local elastic-plastic deformation defect is limited to the angle $2\varphi_0(z)$, the measure of which is correlated with the defect width $c_p(z), 2\varphi_0(z) = \frac{c_p(z)}{R}$ [6–8].

A better modeling of the profile geometry of local plastic deformation defects is obtained if an expression corresponding to a Gauss curve (in polar coordinates) is used in Eq. (2):

$$r(\varphi, z) = R_e - d_p(z)\,exp\left[-\frac{1}{2}\left(\frac{z}{z_0(z)}\right)^2\right]\left[-\frac{1}{2}\left(\frac{\varphi}{\varphi_0(z)}\right)^2\right] \quad (2)$$

where the magnitude of the deviation is given by the term $d_p(z)$, the circumferential extension of the local deformation type defect is limited to the angle $2\varphi_0(z)$, the measure of which is correlated with the width of the defect $c_p(z)$, $2\varphi_0(z) = \frac{c_p(z)}{R}$, and the axial extension variation is given by the term z limited to $\pm S_p/2$, S_p represents the axial length of the anomaly.

To describe with this analytical expression, the shape of the cross sections of a pipeline in the area of an indentation, a computer application, PROFIL_DEF, was created using the Mathcad computing environment. Results obtained by using this computer product and which correctly and objectively describe, from an analytical point of view, the geometric profile of an imperfection or local deformation type defect, are presented in **Figure 6**.

Figure 7 shows a semi-model of the geometric profile corresponding to an imperfection and defect of the local deformation type, made by considering several dimensions of the connection radius that the geometry of the anomaly presents in relation to the geometry of the pipelines that thus present anomalies. The geometric profile was

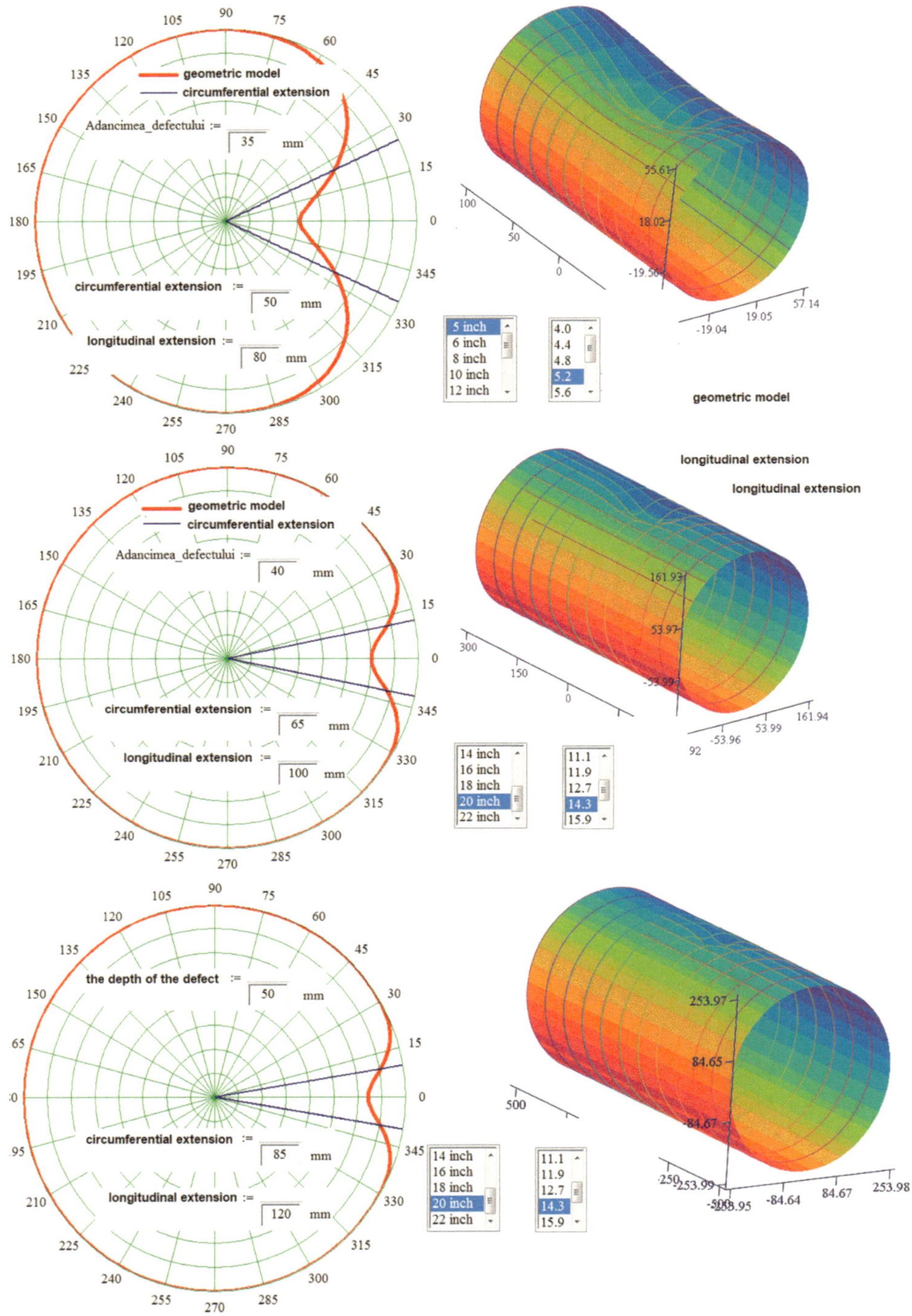

Figure 6.
Results obtained when describing the geometric profile of an imperfection or local deformation type defect.

divided into three contact zones corresponding to the three regions of the profile geometry (AB, BC, and CD).

Two examples of modeling the geometric profile of imperfections and local deformation type defects are also presented in **Figure 7**, considering the parameters and the

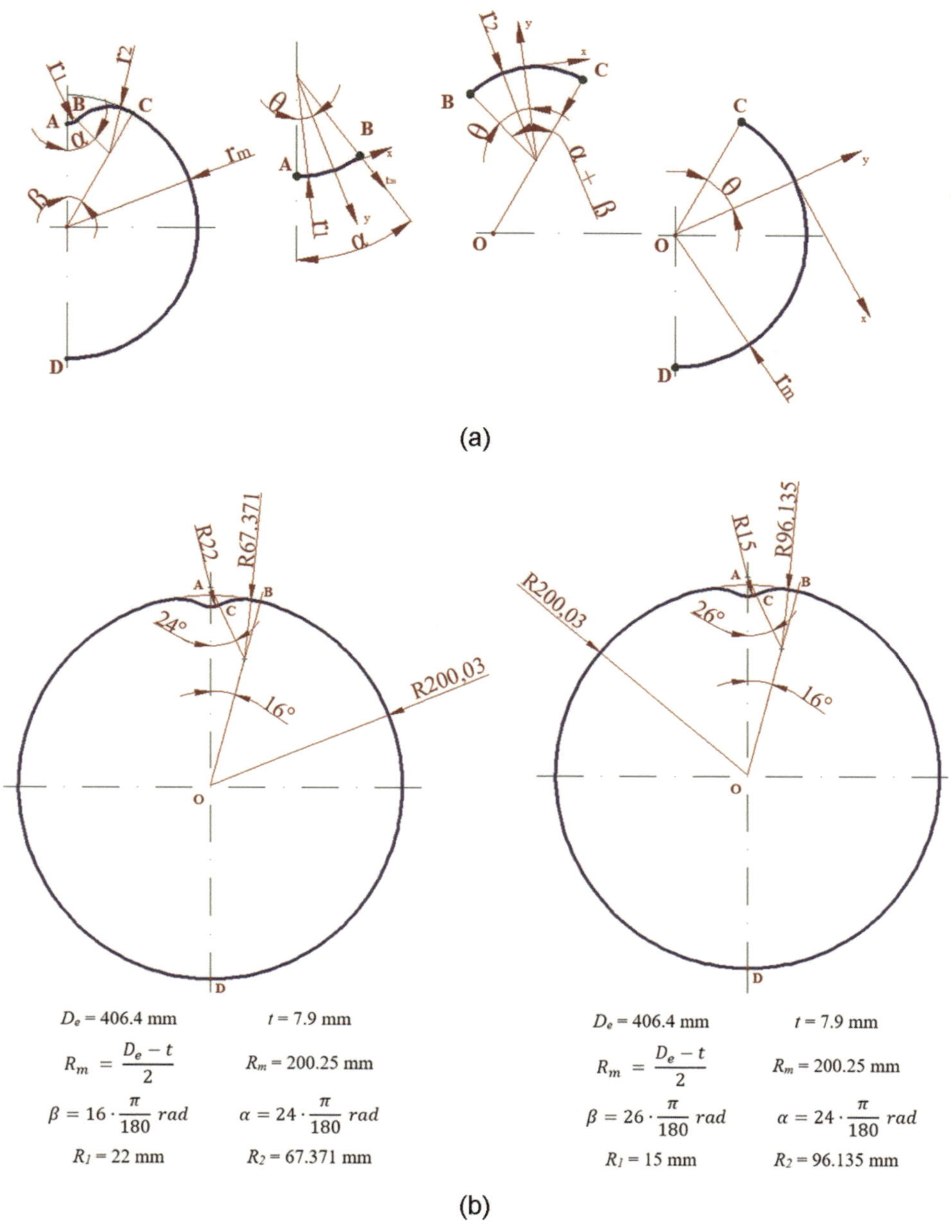

Figure 7.
The general semi-profile of the geometry of an imperfection and local deformation type defect; a–AB zone, b–BC zone, c–CD zone. (a) a - semi-model of the geometric profile for elastic-plastic anomaly, r_1 - main deformation radius in the contact area, r_2- the secondary deformation radius in the contact area, α - main angle corresponding to radius r_1, β - the secondary angle corresponding to radius r_2, rm-the average radius of the pipeline. (b) case study.

method described above, with the observation that the method presented involves a rigorous and difficult determination in practice of the parameters that define defect geometry, modeling that was done using CAD software (AutoCAD).

The verification of the models designed to define the geometric profile of imperfections and local deformation type defects was carried out through experimental tests using ring samples taken from pipes for pipelines. Pipe rings with a diameter of

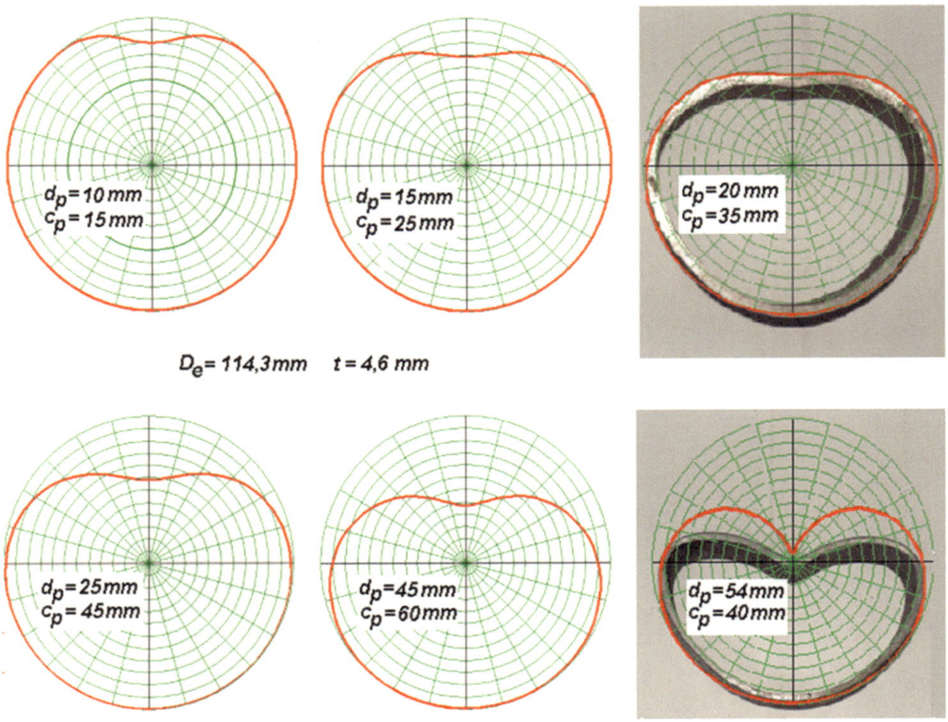

Figure 8.
An example of the application of the proposed model for the analytical description of the configuration of local elastic-plastic deformation anomalies.

De = 114.3 mm, wall thickness *t* = 4.6 mm; 5.4 mm and 8.0 mm and with a length of 50 mm were used for the experimental tests.

The tested pipe section was rested on a prism, and the application of force on the indenter was done gradually with the help of a testing machine, measuring the depth of the anomaly with the help of a dial gauge.

The graphic representations in **Figure 8** summarize the comparisons of the results obtained, depending on the type of indenter used and the wall thickness of the pipe ring used to simulate the appearance of an imperfection and/or local deformation type defect.

From the analysis of the experimental program, the proposed analytical modeling ensures good fidelity for anomalies with relatively small depth, of the type that must be evaluated to be accepted on pipelines.

3. Experimental determination of burst pressure of pipes with anomalies such as local elastic-plastic deformations

The experimental verification of the behavior of pipelines with different types of anomalies under mechanical stress [9, 10] is one of the methods by which their residual bearing capacity is established and leads directly to conclusions regarding the level of confidence that must be associated with the results of the assessment of the severity of the anomalies through analytical methods available. The research carried out concerned the pressure test until bursting of some pipes on which anomalies were made (imperfections and/or anomalies of the type of local elastic-plastic deformations; dents with gouges, considered to be the most dangerous of the anomalies). Two samples were made for the internal pressure test, and each sample consisting of a pipe and two ellipsoidal bottoms welded to its ends, on which three connections were mounted: one for the manometer, one for filling the sample with water and pressurizing to it, and one for venting the sample before pressurization.

The anomalies realized in the two samples subjected to the internal pressure test until breaking/smashing are suggestively presented in **Figure 9**, they are of the indentation type for the first sample, respectively of the indentation type with a gouge for the second sample, and their geometric characteristics are shown in **Table 2**.

The samples that were subjected to the internal pressure test and were made from pipe sections with a diameter of *De* = 163 mm, the wall thickness being *t* = 8 mm. The stand on which the samples were tested is reproduced schematically in **Figure 10**, in

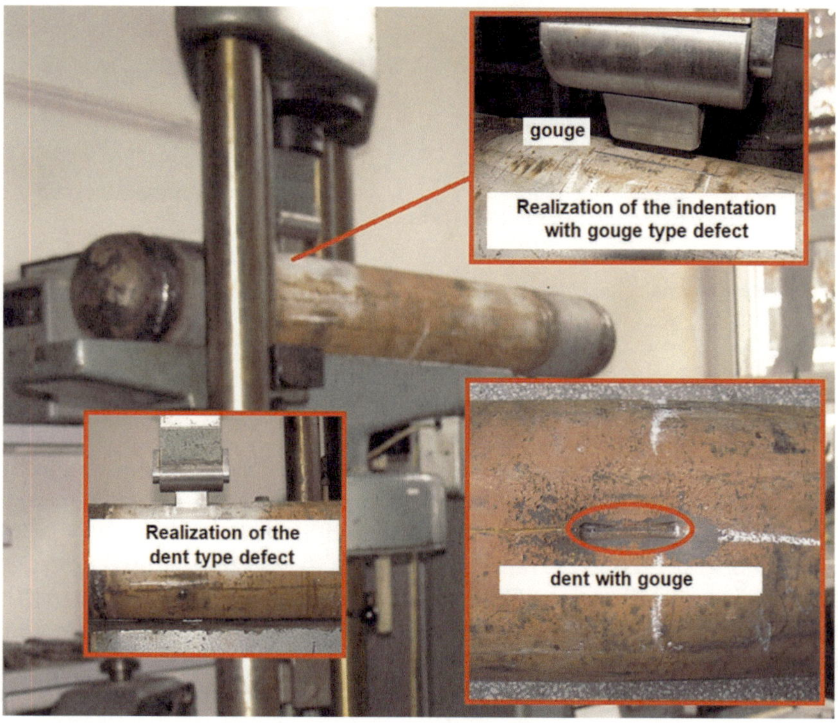

Figure 9.
Realization of the anomalies on the two samples subjected to the test at internal pressure until bursting.

No.	Sample 1	Sample 2		
	Dent	*Gouge*		*Dent*
	Depth [mm]	Depth [mm]	The length on the direction longitudinal [mm]	Depth [mm]
1	25	0,5	40	25
2	30	1,0	40	30

Table 2.
The geometric characteristics of the anomalies made on the samples.

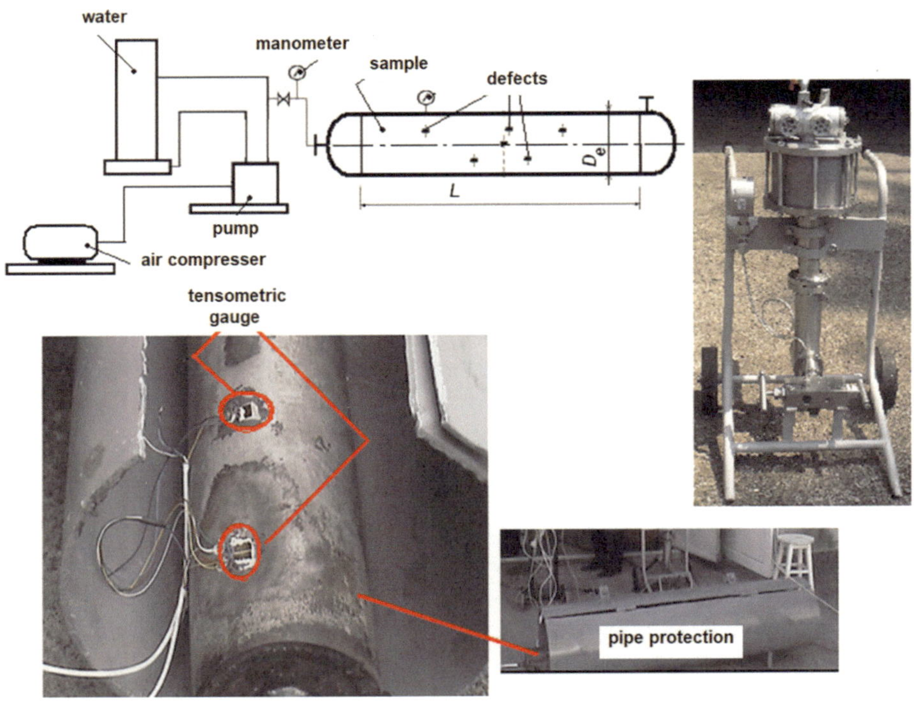

Figure 10.
Scheme and main components of the stand used for internal pressure testing of pipe samples with local surface anomalies.

which the constructive elements of the high-pressure pump and the work platform from the composition of this stand are presented.

When carrying out the experimental research on the samples subjected to the tests, four strain-resistive transducers were applied around the anomaly, two in the circumferential direction (TER 1 and TER 3) and two in the axial direction (TER 2 and TER4) and two strain-resistive transducers, one in the circumferential direction (TER 5) and one in the axial direction (TER 6). During test, the computer controls the acquisition of data with the help of the SPIDER 8 device through the dedicated software CATMAN.

The method of carrying out the tests and the results obtained (quantitative and qualitative), regarding the behavior of the samples during the tests, are presented below:

A. The sample from Pipe 1 resisted up to the pressure 120 bar when failure occurred through cracking of pipe near the joint between the caps and the pipe, outside the defect—see **Figure 11**;

B. The sample from Pipe 2 withstood up to a pressure of 225 bar, when the sample failed by breaking next to the cavity indentation anomaly—see **Figure 12**.

The processing of the experimental results (see **Figure 13**) was carried out by determining the mechanical stresses in the circumferential direction and in the axial direction, using the known formulas:

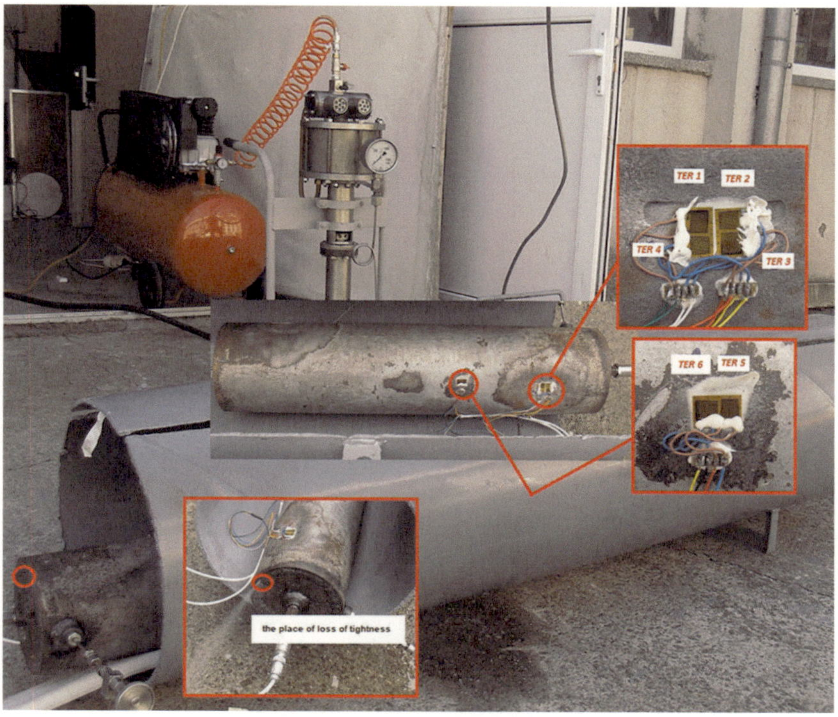

Figure 11.
Images regarding the behavior of sample 1 in the internal pressure test.

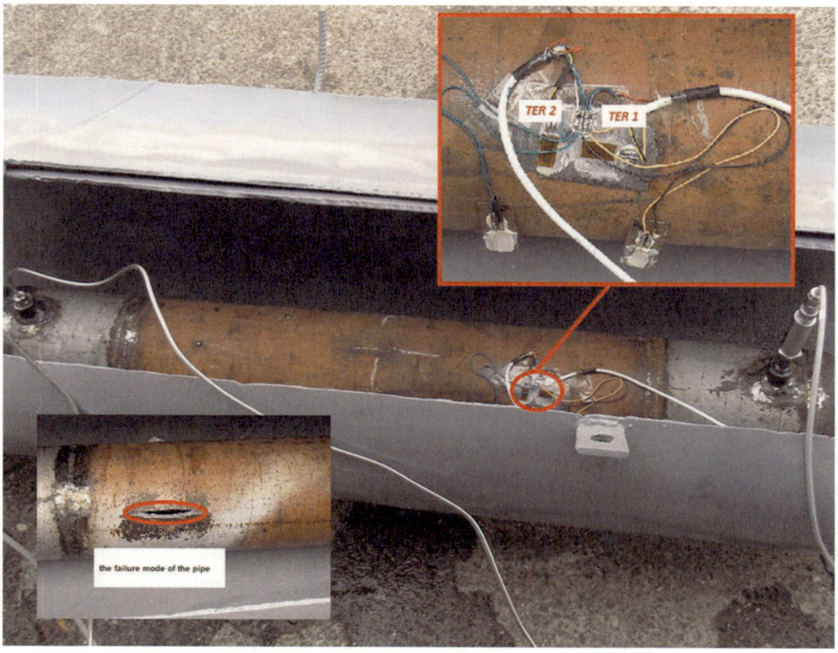

Figure 12.
Images regarding the area where the burst occurred for sample 2.

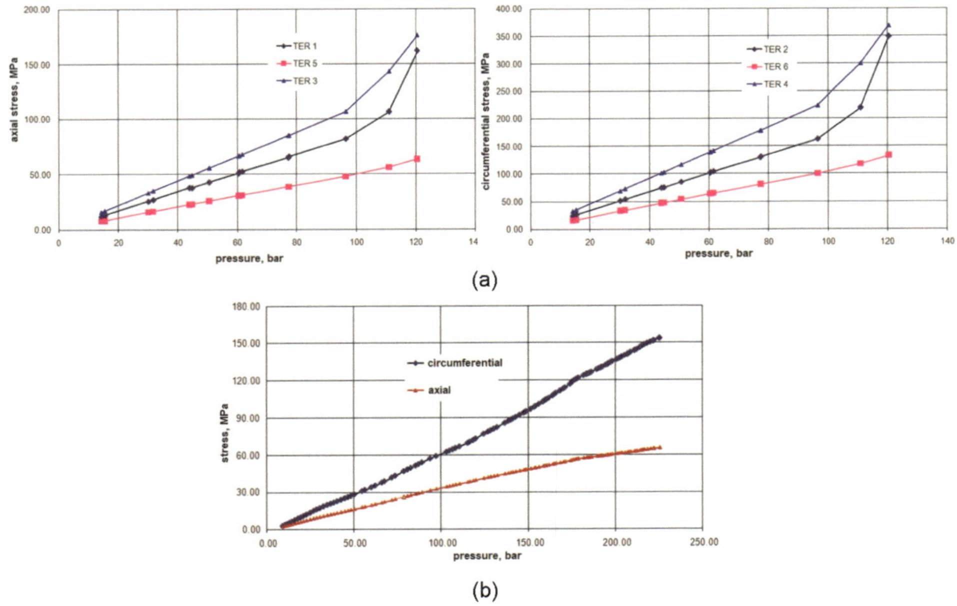

Figure 13.
The results of the experimental analysis by the resistive tensometry method. a. results for sample, b. results for sample 2.

$$\sigma_{\theta ij} = \frac{E}{1-\mu^2}\left[\varepsilon_{\theta i} + \mu\varepsilon_{zj}\right]; \sigma_{zij} = \frac{E}{1-\mu^2}\left[\varepsilon_{zj} + \mu\varepsilon_{\theta i}\right], \tag{3}$$

where E represents the longitudinal modulus of elasticity, and μ—Poisson's coefficient for the sample steel; the specific deformations in the circumferential direction $\varepsilon_{\theta i}$ or in the axial direction ε_{zj} (i and j being the identification numbers of the transducers).

For the determinations, tensor-resistive transducers were used, with a base of 10 mm and resistance 120 Ω. With each of these transducers, at different pressures, during the test of the sample, the specific deformations in the circumferential direction and in the axial direction were determined.

4. Conclusions

The main causes that determine the degradation of pipelines, usually installed underground, are corrosion, third-party interventions and manufacturing, and construction defects; in the last period, there was a significant increase in concessions produced by third-party interventions/interferences.

The principles of codification of imperfections and defects of different types and categories ensure the synthetic and, at the same time, comprehensive specification in the technical documentation regarding the operation, exploitation, maintenance of pipelines of any anomalies found, and their causes; also, the coding of imperfections and defects is particularly useful for the development of maintenance procedures for pipelines.

Anomalies of the elastic-plastic deformation type may have the character of dents (deviations from the circular shape of the cross section of the pipe tubing, obtained by local deformation of the pipe, inward, without removing the material and, as a result, without reducing the wall thickness), gouges (areas of pipe where the wall thickness has been locally reduced due to material removal by mechanical action) or pitted indentations (indentations that have gouges at the bottom of the deformed zone).

Modeling the geometry of anomalies produced by local elastic-plastic deformation is particularly important because it serves to develop evaluation procedures and characterize the severity of these anomalies.

Experimental research into the operational behavior of pipelines with anomalies is usually carried out on stands that allow pressure testing of pipe or pipe sections, with or without anomalies. Usually, these stands can ensure the determination of the stress states of the tested samples and the pressure at which the respective pipe samples fail.

The stand designed and built allows the investigation of the behavior of pipes with or without anomalies, being able to provide both results obtained with the help of electro-tensometry transducers applied to the sample and the bursting pressure of the sample.

Author details

Alin Dinita
Petroleum-Gas University of Ploiesti, Ploiesti, Romania

*Address all correspondence to: adinita@upg-ploiesti.ro

References

[1] Alexander CR. Review of experimental and analytical investigations of dented pipelines. In: Pressure Vessel and Piping Conference. Boston, Massachusetts: 1999. Available from: https://www.chrisalexander.com/wp-content/uploads/2020/05/5.pdf

[2] Iflefel IB, Moffat DG, Mistry J. The interaction of pressure and bending on a dented pipe. International Journal of Pressure Vessels and Piping, 2005;**82**(10):761-769. ISSN 0308-0161

[3] Kim JW, Park CY. Experimental investigation of the failure behavior of notched wall-thinned pipes. Nuclear Engineering and Design. 2006;**236**(17):1838-1846. ISSN 0029-5493

[4] Alexander CR, Kiefner JF. Effects of Smooth and Rock Dents on Liquid Petroleum Pipelines. In: API Pipeline Conference, Dallas, Texas. 1998. Available from: https://www.chrisalexander.com/wp-content/uploads/2020/05/7.pdf

[5] Rinehart AJ. Effects of Localized Geometric Imperfections on the Stress Behavior of Pressurized Cylindrical Shells. Office of Graduate Studies of Texas A&M University; 2003. Available from: https://core.ac.uk/download/pdf/147122554.pdf

[6] Roovers P, Bood R, Galli M, Marewski U, Steiner M. EPRG methods for assessing the tolerance and resistance of pipelines to external damage. In: Denys R, editor. Pipeline Technology Proceedings of the Third International Pipeline Technology Conference. Amsterdam: Elsevier; 2000. pp. 405-425

[7] Rosenfeld MJ. Investigations of dent rerounding behaviour. In: Proceedings of Second International Pipeline Conference, IPC-98. Calgary, Canada: American Society of Mechanical Engineers; 1998. pp. 299-307

[8] Allouti M, Schmitt C, Pluvinage G. Assessment of a gouge and dent defect in a pipeline by a combined criterion. Engineering Failure Analysis Volume. 2014;**36**:1-13

[9] Zhaoa P, Shuaia J, Sunb M, Lva Z, Xua K, Yu W. Burst pressure of thin-walled pipes with dent and gouge defects. Thin-Walled Structures. 2021;**159**:107213

[10] Wanga J, Shuaib Y, Fengc C, Zhanga P, Wangd T, Lina N, et al. Multi-dimensional mechanical response of multiple longitudinally aligned dents on pipelines and its effect on pipe integrity. Thin-Walled Structures. 2021;**166**:108020

Section 2

Pipeline Design and Management

Chapter 5

Flexible Pipes

Minggang Tang

Abstract

An unbonded flexible pipe is one of the most important equipment in offshore engineering, transporting oil and gas between the floater on the sea and the well located on the seabed. Flexible pipes consist of several metallic helical-reinforced layers and internal and external polymer sheaths, and relative slip between the layers is allowed, so that the structure show high axial stiffness and radial stiffness associated with relatively low bending stiffness. During the operation and installation, the flexible pipe will be subjected to complex and coupled loads such as tension, internal pressure, external pressure, torsion and bending, which lead to multiple structural failures. This chapter will present the current theoretical models and research progress to effectively evaluate the response of such composite structure, providing reference ideas for the engineering design of the flexible pipes.

Keywords: offshore flexible pipe, helical wire, complex loads, interlayered pressure, structural response

1. Introduction

Pipelines are the "lifeline" equipment for the production and development of offshore oil and gas resources and mineral resources. The traditional steel pipeline technology is mature, but the limitations in bendability, corrosion protection, installation and laying speed have significantly increased its comprehensive cost, so the concept of offshore flexible pipeline came into being. It can be bent into a small bending radius without reducing the mechanical bearing capacity, and can be installed and recycled by the reel method, so the comprehensive cost is relatively low⊠Meanwhile, in some medium water depth environments, using offshore flexible pipes as risers is the only solution. Therefore, since the 80s of last century, flexible pipelines have gradually been paid attention to and applied in the development of offshore oil and gas resources, and flexible pipelines account for more than 80% of the marine risers currently in service [1].

Each reinforcement layer is specifically designed to accommodate a specific load. Under the axisymmetric loads, such as tension, internal pressure, external pressure, and torsion, the bearing capability of the reinforced layer usually depends not only on the structural design of the layer itself, but also on the interaction with the adjacent layers [2], so that the pipe wall is formed by the armor layers. While, under the combined non-axisymmetric loads such as bending loads, the interlayered interaction becomes more complicated and makes the mechanical behavior of the reinforced components show the strong nonlinearity. These bring challenges into accurately evaluating the structural strength and fatigue performance of flexible pipes.

The objective of the current study is to comprehensively establish the latest theoretical models of structural responses of offshore flexible pipes under various typical loads, and analyze the coupling relationship between the structural deformation and interlayered interaction under different loads. Especially for the tension armor wires, the calculation method of nonlinear responses under combined tension and bending is developed, and the theoretical models are verified by one case. These can not only help readers clearly understand the bearing principles and the response rules of flexible pipelines, but also provide a comprehensive, convenient and effective tool for carrying out the quantitative design and evaluation of offshore flexible pipeline in practical engineering.

This chapter is organized as follows. Section 2 puts forward the design requirements of flexible pipelines from the perspective of engineering applications; Section 3 describes the mechanical principles and structural types of flexible pipes; Section 4 introduces theoretical models of pipeline structural responses for axisymmetric loads such as tension, internal pressure, external pressure and torsion; The mechanical responses of pipes under non-axisymmetric loads such as bending and complex loads are discussed in Section 5; The last section takes one 8 inches internal diameter pipeline in practical engineering as example to quantitatively describe the mechanical responses of flexible pipeline structure under various loads.

2. Offshore flexible pipe design requirements

1. On the basis of ensuring normal flow, the design of offshore flexible pipelines should first achieve better bendability, which means compliance with strenuous movement of the floater in high sea conditions, as well as rapid manufacturing, laying and recovery by the reel method. The flexible pipe with an inner diameter of 8 inches can be rolled on a drum with a radius of 2 meters for storage, and the speed of the reeling method is about 7 times that of conventional steel pipe installation.

2. Flexible pipelines need to ensure that the internal high temperature (up to 130°C), the complex fluid medium does not leak outward, and the external seawater does not erode inward, so as to form a complete pipe wall space and protect the internal metal components from corrosion.

3. In order to transport the medium from the seabed of thousands of meters to the sea surface, the pipeline needs to have a high resistance to the continuous action of internal pressure (the general working pressure of a 4-inch inner diameter pipe is 25Mpa).

4. As the water depth gradually increases, the pipeline body will be subjected to huge external pressure. In order to prevent local or global buckling collapse failure, the flexible pipe wall needs to be able to continuously resist the external pressure from seawater (about 30Mpa at a depth of 3000 meters).

5. During the installation and operation of flexible risers (as shown in **Figures 1** and **2**), the upper end of the pipeline is suspended on the floater on the sea surface [3]. Due to the self-weight, the flexible riser close to the floater will be

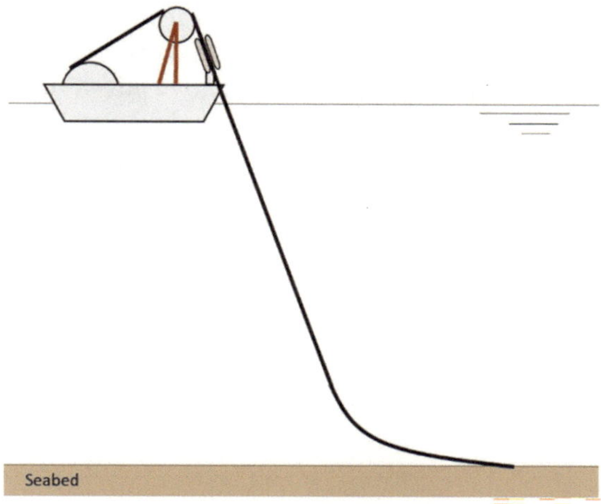

Figure 1.
Schematic diagram of the flexible pipe laying- J lay.

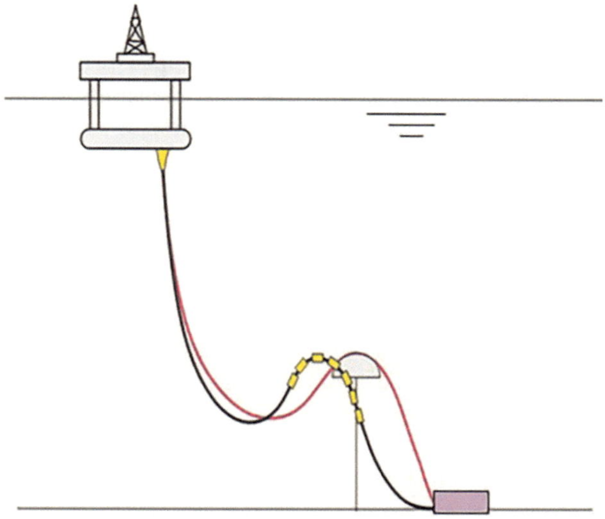

Figure 2.
Schematic diagram of flexible riser in operation.

subjected to huge axial tension (when the flexible riser of an 8-inch inner diameter is used in water depth of 1500 meters, the axial tension can generally reach more than 100 tons). Therefore, the pipeline should have the low bending stiffness associated with the high axial stiffness and radial stiffness.

6. During the installation using reeling method, flexible pipes are subjected to repeated axial torsion. And during normal operation, the pipeline itself will also twist due to the movement of the floater and dynamic environment. Therefore, the torsion resistance in the clockwise/counterclockwise direction of flexible pipes needs to be considered during the design process, and the torsional equilibrium is also required, i.e. the pipeline does not twist freely under other loads.

7. Owing to its self-weight coupled with the floater motion and dynamic environment, the flexible riser at the interface with the floater (generally within 50 m) will encounter relatively alternative curvatures as well as high axial loads and internal pressure in operation, and stress fatigue failure is more likely to occur on the tension armor layer in the pipeline [4]. Currently, the service life of Flexible pipes is designed to reach 25 years at least, so accurately assessing the fatigue resistance of pipelines is a constant need for oil companies and owners.

In summary, the flexible pipeline structure needs to have excellent bending performance and corrosion resistance, and at the same time, it must have good resistance in internal pressure, external pressure, tension, torsion and fatigue. That is, while giving full play to the advantages of flexibility, it can meet the safety and reliability of the structure.

3. Mechanical principle and structural type of offshore flexible pipeline

3.1 Mechanical principle

According to the design requirements of flexible pipes, it is no longer feasible to use homogeneous materials for the entire pipe wall. Considering the need to seal inside and outside the pipe, the innermost and outermost layers of the pipe are usually made of polymer materials of a certain thickness. In order to meet the bearing capacity needs in all directions at the same time, the metal reinforcement structure in the form of a spiral is required. Considering that the single-layer helical structure is difficult to balance and achieve the design resistances, it is necessary to increase the spiral reinforcement layer according to different bearing capacity requirements, and maintain an unbonded form between the layers to reduce the minimum bending radius of the pipeline. The following describes the strengthening principle of pipeline structure according to different load resistance requirements.

1. In order to withstand the internal pressure, the outside of the innermost polymer material needs to be spiral strengthened, and the helical winding angle should reach more than 85° (the angle to the axis of the pipeline). At the same time, in order to ensure the stability of the pipe body section during the dynamic operation, the adjacent spiral members usually adopt a lap structure, so the pressure armor layer is usually made of metal components with special-shaped sections spiral wound at a large angle, as shown in **Figure 3**. Depending on the amount of internal pressure, the cross-sectional shape, thickness and winding angle of the wire can be specially designed.

2. In order to withstand high external pressure, spiral reinforcement is usually required on the innermost side of the pipe. This layer is similar to the pressure-resistant armor layer, which needs to be wound at a large angle. Considering that the layer is directly washed by the internal transport fluid medium, and needs to have higher structural stability and lumen smoothness, the carcass layer usually adopts the form of interlocking structure. By changing the shape, thickness, angle and other parameters of the cross-section of the layer component, the bearing capabilities to different grades of external pressure can be realized.

Figure 3.
Typical cross-section of the pressure armour layer.

3. The individual helical wire is similar to a spring. Under the axial tension, the bending moment and torsion exist on the cross section without the tensile force along the helical direction, so the spring resistance to the axial tension is very limited. And the larger the helical angle, the smaller the cross-sectional moment and the lower the resistance to axial tension. When the helical wire is supported by a cylindrical component inside, the tensile force along the helical direction will dominate under the axial tension [5] and as the helical angle reduces, the resistance to the axial tension improves significantly. Therefore, the tension armor layer is usually formed by many steel wires with rectangular cross sections winding at an angle between 30 degrees to 60 degrees on the internal pressure armor layer. According to different tension loads, the wire cross section size, the number of wire and the helical angle can be specially designed.

4. The demand on the helical components for the torsional resistance of pipeline is similar to that for the tension resistance, so the tension armor layer also provides torsional resistance. In order to achieve the torsional balance [6], the pipe is usually provided with two tension armor layers, and the inner and outer tension armor layers are spirally wound at opposite angles.

It can be seen that in addition to the inner and outer polymer sheaths, the flexible pipe can simultaneously realize the overall flexibility and the resistance to loads in other directions by setting different types of metal spirals in the annulus.

3.2 Typical structure

According to the mechanical principle, the typical structure of the offshore flexible pipeline used in deep water is shown in **Figure 4**. From the inside to the outside, they are respectively:

- Carcass layer: interlocked section wire helically wound, mainly used to resist uniform external pressure from seawater, and avoid the erosion of the inner

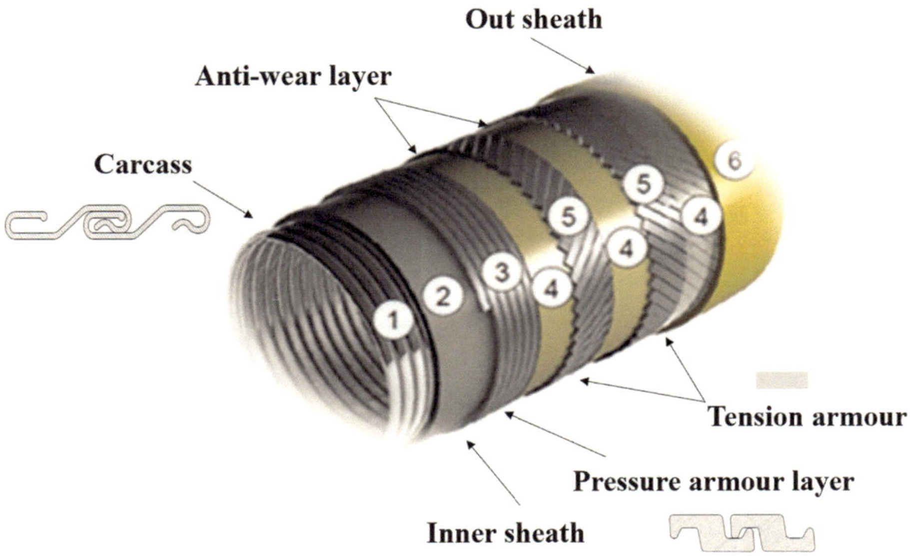

Figure 4.
Schematic diagram of typical structure of flexible pipeline.

sheath from the medium. The helical angle between the direction of the wire and the axial direction of the pipeline is close to 90 degrees, and the material is usually stainless steel to avoid corrosion caused by the internal medium.

- Inner sheath: cylindrical polymer sheath formed by continuous extrusion, and mainly used to seal the internal medium while transferring internal pressure. Its common materials are: HDPE (high-density polyethylene), XLPE (cross-linked polyethylene), PA (nylon series), PVDF (polyvinylidene fluoride) and so on. Different materials have different mechanical properties, permissible temperature and permeability, and need to be selected according to the types of the internal medium. Among them, the permissible temperature of PVDF can reach 130 degrees Celsius, which is the upper limit of the current design temperature of flexible risers.

- Pressure armor layer: special-shaped section (such as "Z" type, "C" type and "T" type, etc.) steel wire helically wound, mainly used to resist the internal pressure from the inner sheath, and also play a certain role in resistance to the external pressure. The helical angle of this layer is close to 90°, and the material is usually pipeline steel or special carbon steel, which yield strength is above 800Mpa.

- Tension armor layer: rectangular section steel wire helically wound, mainly used to resist the axial tension and ensure the torsional balance of the pipeline itself. The helical angle of this layer is usually between 30 degrees and 60 degrees, and the material is usually pipeline steel or special carbon steel, which yield strength is also above 800Mpa.

- Anti-wear layer: the polymer tape with a very small friction coefficient helically wound, mainly used to reduce the possible friction and wear between the metal layers without bearing the loads.

- Outer sheath: cylindrical polymer sheath formed by continuous extrusion, and mainly used to seal the external sea water while transferring the external pressure. Its common materials are HDPE, PA, etc.

In addition, according to the specific needs from users, layers can be added or subtracted on the basis of the above basic structure. The composite pipe wall is unbonded, and the whole is formed through the interaction between the layers, so it is of great significance to accurately understand the structural responses under different loads for designing and evaluating the safety of the pipeline.

4. Mechanical behavior of pipelines under symmetrical loads

According to the symmetry of the load on the pipe relative to the central axis, the loads including axial tension, internal pressure, external pressure and torsion are collectively referred to as the axisymmetric loads. Under this type of loads, the adjacent layers in the pipe wall contact and squeeze with each other.

4.1 Resistance to internal pressure

Flexible pipes are usually designed using the pressure armor layer to independently bear the internal pressure. The internal pressure is transferred through the inner sheath layer to the pressure armor layer. Considering that the cross-sectional area of the armor wire has a direct impact on its internal pressure resistance, the special-shaped section can be simplified to a rectangular section with the same thickness and material and certain voids in the design stage. Based on the assumption of line elasticity and small deformation, an analytical model is established for the general mechanical behavior of the spiral steel strip with rectangular section under internal pressure, in which the axial stress along the direction of the steel wire is concerned.

In case of the helical angle of pressure armor wire 90°, a closed plane ring is acted with radial pressure P along circumference of cylinder core, as shown in **Figure 5a**.

Referring to the most flexible pipe structures, the thickness of wires is much smaller than the diameter of the internal core. Then the classic Lame formula of the plane problem [7] can be used and equilibrium can be written as:

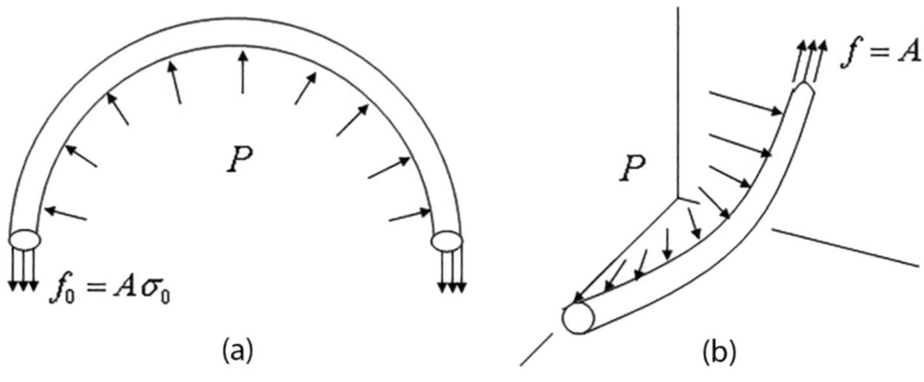

Figure 5.
Helical steel wire with radial pressure- (a) helical angle is 90°, (b) helical angle is not 90°.

$$P = \frac{\sigma_0 h_0}{R} \quad (1)$$

In which, R means the helical radius of the pressure armor layer, σ_0 denotes the circumferential stress, that is the axial stress of the wire, h_0 is the equivalent thickness of the wire.

In case of the helical angle of steel wire not 90°, that is, the helical wire is pressured in the radial direction illustrated in **Figure 5b**. Let f be the axial force along the wire and σ represents axial stress, the component force in hoop direction from the axial force can be expressed as $f_0 = f \times \sin\alpha$. And the mapping sectional area in hoop direction of the wire is $a = A/\sin\alpha$. Then the stress of the wire in hoop direction is:

$$\sigma = \frac{f}{A} = \frac{f_0/\sin\alpha}{a\sin\alpha} = \frac{f_0}{a\sin^2\alpha} = \frac{\sigma_0}{\sin^2\alpha} \quad (2)$$

Substituting Eq. (2) into (1), the relationship between radial pressure from single armor layer and axial strain along the wire can be obtained as [8]:

$$P = \frac{h_0 \sigma \sin^2\alpha}{R} \quad (3)$$

It can be seen that as the internal pressure increases, the tensile stress along the direction of the wire axis increases. Considering the thickness, helical angle and radial radius of the wire unchanged, the stress along the wire direction and internal pressure basically show a linear relationship.

4.2 Resistance to external pressure

Flexible pipes are usually designed using the innermost carcass layer to independently bear the external pressure. Although the sea water pressure generally acts on the outer sheath, the outer sheath leaks in extreme cases and the seawater passes through the annulus and directly acts on the outside of the inner sheath. In the design phase, the interlocked carcass layer can be equivalent to a homogeneous cylinder with a certain thickness. Then the problem can be simplified to analyze the buckling collapse of the cylinder under external pressure. This section establishes an analytical model of the general mechanical behavior of a homogeneous ring (cylinder) under external pressure based on the theory of elastic stability.

Timoshenko and Gere [9] first gave the deflection of a thin bar with a circular cross-section as shown in Eq. (4), for the problem of the elastic buckling of a ring or tube. It was assumed that radial displacement is small and the displacement in the tangential direction can be ignored.

$$\frac{d^2\omega}{d\theta^2} + \omega = -\frac{1}{EI} q R^3 (\omega + \omega_1 \cos 2\theta) \quad (4)$$

where ω denotes radial deformation of a thin bar, θ is the angle in the hoop direction, M is the bending moment loading on the bar, EI is the bending stiffness of the bar cross section, R is the mean radius of curvature, ω_1 is the maximum initial radial deviation, and q means a uniform external pressure on the plane ring.

Considering the continuity condition on points A, B, C and D on the plane ring in **Figure 6**, the analytical solution of Eq. (4) is:

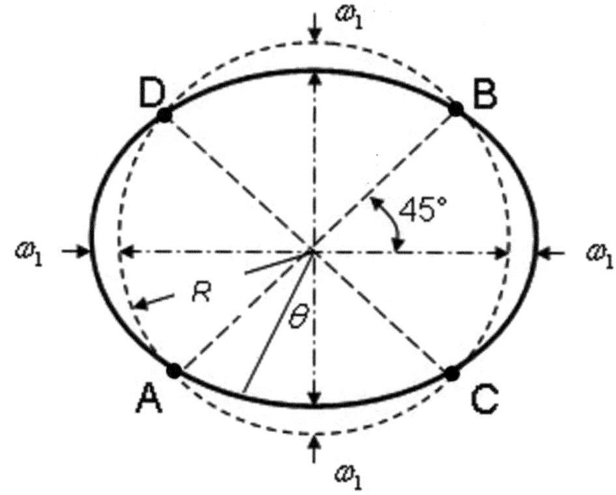

Figure 6.
Schematic diagram of a plane ring with an initial imperfection (ovality).

$$\omega = \frac{\omega_1 q}{q_{cr} - q} \cos 2\theta \quad (5)$$

Eq. (5) gives the analytical expression for the elastic buckling of a plane ring under uniform external pressure, where the critical pressure q_{cr} of the ring is given by:

$$q_{cr} = \frac{3EI}{R^3} \quad (6)$$

In the case of a rectangular ring section and the plane strain condition, the inertia moment I per unit length can be written as $t^3/12$ and E is converted to $E/(1-\nu^2)$. Then the critical pressure value under the plane strain conditions can be expressed as follows:

$$q_{cr} = \frac{E}{4(1-\nu^2)} \left(\frac{t}{R}\right)^3 \quad (7)$$

where t represents the equivalent thickness of the carcass layer resulted from equivalent methods. It can be seen that the critical pressure and the equivalent thickness of the carcass layer basically show a cubic relationship. It should be noted that in the design stage, it is critical to use conservative and effective equivalent methods to determine thickness. Current equivalent methods include area equivalence [10], bending stiffness equivalence per unit length [11], bending stiffness equivalence per unit area and strain energy equivalence methods [12]. Through the experimental verification, the equivalent thickness obtained by the strain energy equivalent method can keep the critical pressure in a conservative state, which is convenient for engineering applications.

4.3 Resistance to axial tension

The double tension armor layers of flexible pipes are designed mainly to resist the axial tension, meanwhile the structure of internal core is designed to provide a support

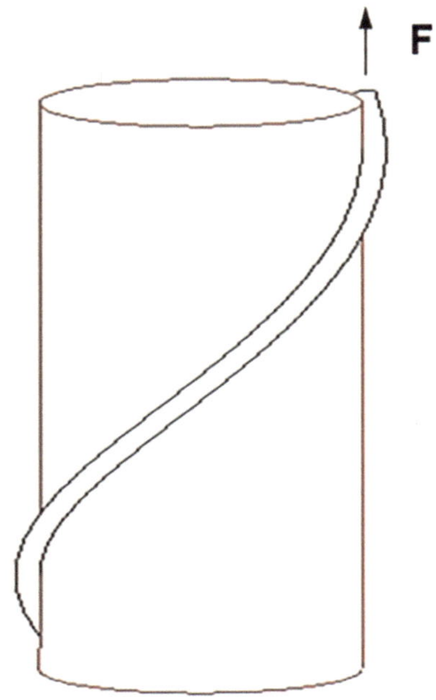

Figure 7.
Mechanical model for tensile stiffness of flexible pipeline.

in the radial direction. To capture structural character, a simplified model that a column is wound by helical steel wires is established to analyze the tensile stiffness as illustrated in **Figure 7**.

To simplify the deducing of the tensile stiffness, the following assumptions are made in the model: (i) Helical wires in any layer are equally spaced around the circumference of the flexible pipe. (ii) Only the axial deformation of the wires is considered during tension and the bend stiffness and torsional stiffness of wires are neglected. (iii) The internal core is modeled as a cylinder which has radial deformation under pressure from armor wires. (iv) All elements meet the assumption of small deformation and all possible frictions are neglected. (v) The helical wires and cylinder core element are homogeneous, isotropic and linearly elastic.

The analysis will be carried out on the plane for convenience by unfolding the cylinder with a helical pitch length as shown in **Figure 8a**. Considering axial tension F on the pipe model and the axial strain ε, we analyze the two situations below [13].

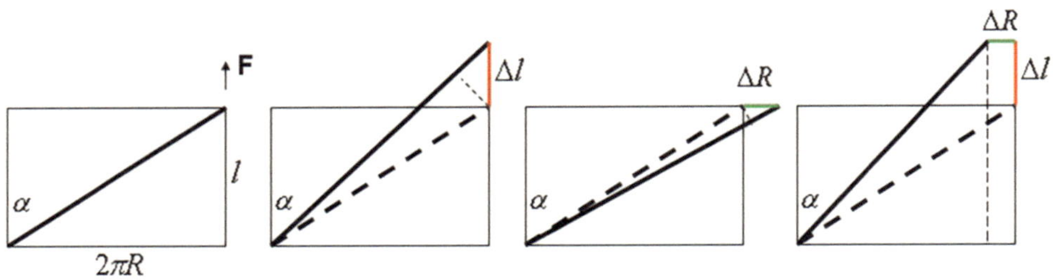

Figure 8.
Deformation of steel wire under axial tension.

1. If only the axial deformation of the pipeline is considered, Δl is defined as axial deformation just as shown in **Figure 8b**. According to the geometrical relationship, the elongation along wire axial direction can be expressed as $\Delta l \cdot \cos\alpha$, in which α means the angle between umbilical axial direction and wire axial direction. Considering the length of the helix wire before deformation $l_0 = l/\cos\alpha$, we can obtain the strain in the wire axial directions as follows:

$$\varepsilon = \frac{\Delta l \cdot \cos\alpha}{l_0} = \frac{\Delta l}{l}\cos^2\alpha \tag{8}$$

2. If we only consider the radial deformation of pipeline and let ΔR be radial elongation as shown in **Figure 8c**, the elongation along wire axial direction is $\Delta R \cdot \sin\alpha$. If we define the length of the helix wire before deformation $R_0 = R/\sin\alpha$, where R means radius of the helix wires, the strain of wire can be expressed as eq. (9).

$$\varepsilon = \frac{\Delta R \cdot \sin\alpha}{R_0} = \frac{\Delta R}{R}\sin^2\alpha \tag{9}$$

However, the actual deformation is the synthetization of the axial deformation from tension and the radial deformation from the contraction of cylinder core caused by the pressure of the outer tension armor layers as shown in **Figure 8d**. Add the eq. (8) and (9), the synthetization strain can be written as:

$$\varepsilon = \frac{\Delta l}{l}\cos^2\alpha - \frac{\Delta R}{R}\sin^2\alpha \tag{10}$$

From eq. (10), the axial tension of all the armor wires can be expressed as:

$$\begin{aligned} F &= \sum_{i=1}^{m} EA\varepsilon_i \cos\alpha_i \\ &= \frac{\Delta l}{l}\left(\sum_{i=1}^{m} EA\cos^3\alpha_i\right) - \Delta R\left(\sum_{i=1}^{m}\frac{EA\sin^2\alpha_i\cos\alpha_i}{R_i}\right) \\ &= \Theta_1\frac{\Delta l}{l} - \Theta_2\Delta R \end{aligned} \tag{11}$$

where E is Young modulus of steel wire, A is cross-sectional area of the wire, m identifies the amount of armor layers and i means layer number, n_i is the wire number in ith layer, Θ_1 and Θ_2 are the representative symbols of the corresponding analytical algebraic formulas.

From eq. (3), the relationship between radial pressure from single armor layer and axial strain along the wire can be extended as [14]:

$$P = \frac{nEA\varepsilon\sin^2\alpha}{2\pi R^2\cos\alpha} \tag{12}$$

There are usually two or more armor layers for common flexible pipes. Then the radial pressure on cylinder core should be added by the pressure from all the armor layers. By substituting (10) into (12), the equilibrium equation about P and ε is further written as:

$$P = \sum_{i=1}^{m} \frac{n_i EA \varepsilon_i \sin^2 \alpha_i}{2\pi R_i^2 \cos \alpha_i}$$
$$= \frac{\Delta l}{l} \left(\sum_{i=1}^{m} \frac{n_i EA \sin^2 \alpha_i \cos \alpha_i}{2\pi R_i^2} \right) - \Delta R \left(\sum_{i=1}^{m} \frac{n_i EA \sin^4 \alpha_i}{2\pi R_i^3 \cos \alpha_i} \right) \tag{13}$$
$$= \Psi_1 \frac{\Delta l}{l} - \Psi_2 \Delta R$$

where Ψ_1 and Ψ_2 are the representative symbols of the corresponding analytical algebraic formulas. When the cylinder core is pressed by armor steel wires, it comes to the radial contraction. In order to describe the mechanical phenomenon, the radial stiffness Ω of the core is introduced and defined as:

$$\Omega = P/\Delta R \tag{14}$$

Eqs. (11), (13) and (14) together form a close equation set (15), which leads to the equilibrium of the steel wires with tensions considering the compressible deformation of the internal core.

$$\begin{cases} F = \Theta_1 \dfrac{\Delta l}{l} + \Theta_2 \Delta R \\ P = \Psi_1 \dfrac{\Delta l}{l} + \Psi_2 \Delta R \\ \Omega = \dfrac{P}{\Delta R} \end{cases} \tag{15}$$

By eliminating P and ΔR, the relation of the tension and axial strain can be obtained as:

$$F = \left(\Theta_1 + \frac{\Theta_2 \Psi_1}{\Omega - \Psi_2} \right) \frac{\Delta l}{l} \tag{16}$$

The relationship between the axial tension and the generated interlayer pressure can be analytically expressed as:

$$F = \frac{\Theta_1 \Omega + \Theta_2 \Psi_1 - \Theta_1 \Psi_2}{\Psi_1 \Omega} P \tag{17}$$

If the radial contraction of the internal cylindrical core is not considered, that is, the radial stiffness Ω tends to infinity, Eq. (17) can be further simplified to obtain the explicit expression of axial tension and radial pressure:

$$P = \frac{\Psi_1}{\Theta_1} F = \frac{n \tan^2 \alpha}{2\pi R^2} F \tag{18}$$

It can be found that as the axial tension increases, both the tensile stress along the direction of the wire and the radial pressure on the internal cylindrical core increases, and the interlayer pressure and axial tension basically show a linear relationship. It is precisely because the radial stiffness provided by the pressure armor layer is relatively large, the axial tensile stiffness of the pipeline is sufficient, and the helical armor wire can better withstand the tension. As the radial stiffness of the internal cylindrical components gradually decreases, the deformation of the helical wire will no longer be a small geometric deformation, and each wire will tend to have the independent spring deformation, therefore the tension resistance will be greatly reduced.

4.4 Resistance to torsion

Flexible pipes are primarily designed to resist torque in the clockwise/counter-clockwise direction through the double tension armor layers [15]. The basic theory and assumptions are the same as those in the previous section, and this section focuses on the relationship among the torque, the radial pressure and the axial stress along the wire helical. It can be seen from the eq. (10) that the circumferential component of the axial force along the wire direction resists the overall torque of the pipeline, so the relationship between the torque on the single tension armor layer of the pipeline and the deformation along the wire direction can be expressed as:

$$Q = nREA\varepsilon \sin\alpha \tag{19}$$

Combined with eq. (12), the analytical relationship between interlayered pressure and torque without consideration of the radial contraction can be written as:

$$P = \frac{tan\alpha}{2\pi R^3} Q \tag{20}$$

It can be noticed that as the torque increases, both the tensile stress along the direction of the wire and the radial pressure on the internal cylindrical core increases, and the interlayer pressure and torque also show a linear relationship. Considering both the axial and radial deformation, the derivation process is similar to the previous section, resulting in a more complete expression for eq. (20). It is worth noted that due to the different helical directions of tension armor layers, the direction of each layer's resistance to torque is different. The neighboring tension armor layers will tend to squeeze or separate with each other due to the different torsion direction.

5. Mechanical behavior of pipelines under complicated loads

When the pipeline is subjected to non-axisymmetric loads, such as bending, the relative sliding between neighboring layers occurs due to the unbonded condition, so that the pipeline has excellent flexibility. When the pipeline is subjected to both non-axisymmetric loads and axisymmetric loads at the same time, the tangential friction is generated between the layers due to radial pressure and relative slippage. Especially for tension armor wire, the stress state becomes very complicated as the pipeline curvature changing, which will firstly cause the fatigue failure.

5.1 Bending performance

When the flexible pipe is only subject to bending, the interlayered interaction is weak, and both the deformation and relative slippage exist for each layer. Since the minimum bending radius of the inner and outer sheath and the interlocked armor layer can be obtained by the classical material mechanics, this section focuses on the mechanical behavior of tension armor wire bending under weak interlayered interaction conditions, which provides a theoretical basis to evaluate the bending performance of pipelines.

One helical wire with a rectangular cross-section wrapped around a cylindrical shell with radius r is considered to represent the initial condition of the tension armor. When the model is bent (the radius of curvature is $\rho = 1/\kappa$), a curve is created on the toroid, as illustrated in **Figure 9a**. The angular coordinate θ located along the torus radius and the arc length coordinate u located along the torus centerline are chosen as parameters of the torus surface. The space vector $\mathbf{R}$ on the surface can be expressed in (θ, u)-coordinates instead of in Cartesian coordinates as [16]:

$$R(u,\theta) = \begin{bmatrix} \left(\frac{1}{\kappa} + r \cdot \cos\theta\right)\cos(\kappa u) - \frac{1}{\kappa} \\ \left(\frac{1}{\kappa} + r \cdot \cos\theta\right)\sin(\kappa u) \\ r \cdot \sin\theta \end{bmatrix} \tag{21}$$

Correspondingly, (x_2, x_3)-rectangular coordinates can be built in the cutting plane perpendicular to the centerline of the wire (see **Figure 9b**), where $\boldsymbol{x_2}$ is perpendicular to the local unit wire normal $\mathbf{n}$ and is located along the rectangular width, and $\boldsymbol{x_3}$ points to the local unit wire binormal $\mathbf{b}$ and is located along the thickness.

When the cylindrical shell is bent with a specified curvature, the initial equilibrium state of the helical wire is broken, and the wire is forced to slip on the torus surface. Correspondingly, the changed space vector of the wire can be expressed as $\mathbf{R}'(\theta', u')$. Based on differential geometry theory, the variation of curvature components of the bending wire can be defined as:

$$\begin{cases} \Delta\kappa_n = \kappa_n - {\kappa_n}^0 \\ \Delta\kappa_b = \kappa_b - {\kappa_b}^0 \\ \Delta\tau = \tau - \tau^0 \end{cases} \tag{22}$$

In which, the stress on the wire cross section caused by the normal curvature $\Delta\kappa_n$ and the binormal curvature $\Delta\kappa_b$ is the positive stress and along the wire direction; the stress on the wire cross section caused by the torsional curvature $\Delta\tau$ is the shear stress

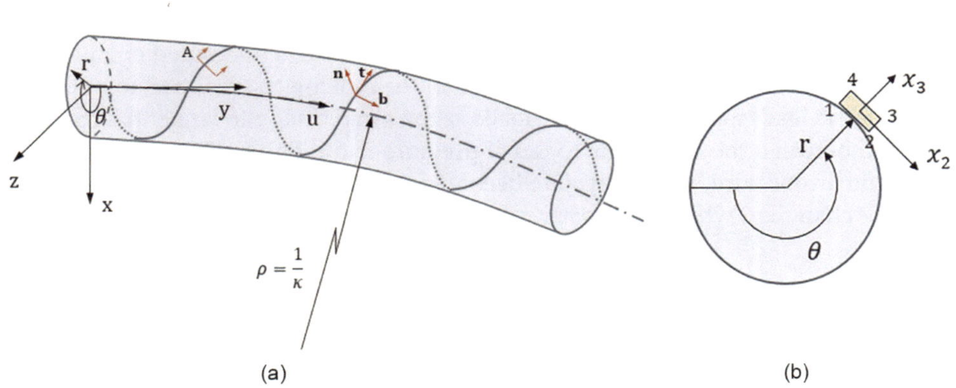

Figure 9.
Wire geometry on a toroid under the curvature- (a) overall view, (b) cross-section A-A view.

and perpendicular to the wire direction. During the wire bending, the fatigue failure usually occurs firstly at the corner point of the wire rectangular section due to high local stress, which needs to be concerned. Since the shear stress at the corner point is zero, the stress is in the uniaxial stress state and can be expressed as:

$$\sigma^{y} = \sigma_{n} + \sigma_{b} = \left(\Delta\kappa_{n} \times x_{2y}\right) \cdot E + \left(\Delta\kappa_{b} \times x_{3y}\right) \cdot E \tag{23}$$

where y represents the corner point number.

Based on the existing experimental results and theoretical models, the bending behavior of the helical wire considering the weak interlayered interaction is closer to that of a spring [4]. Therefore, according to the classical principle of elasticity, the curvature variation in the three directions of the helical wire can be obtained as eq. (24) when the pipeline has a certain curvature κ.

$$\begin{cases} \Delta\kappa_{n} = \dfrac{B\cos\theta}{EI_{n}}\kappa \\ \Delta\kappa_{b} = -\dfrac{B\sin\theta\cos\alpha}{EI_{b}}\kappa \\ \Delta\tau = \dfrac{B\sin\theta\sin\alpha}{GJ}\kappa \end{cases} \tag{24}$$

Use of Eq. (24) in Eq. (23) leads to the stress on a corner point of the helical wire:

$$\sigma^{y} = \frac{B\cos\theta}{EI_{n}}\kappa E \cdot x_{2y} - \frac{B\sin\theta\cos\alpha}{EI_{b}}\kappa E \cdot x_{3y} \tag{25}$$

in which B represents the bending stiffness of the spring and is expressed as $2\cos\alpha / \left(\frac{1}{EI_n} + \frac{\cos^2\alpha}{EI_b} + \frac{\sin^2\alpha}{GJ}\right)$, EI_n and EI_b represent the bending stiffness of the rectangular cross-section with respect to the x_3, x_2 axis, respectively, and GJ indicates the torsion stiffness of wire section.

It can be seen that under the condition of weak interlayered interaction, the bending behavior of flexible pipes can be evaluated layer by layer, and then the minimum bending radius (MBR) can be comprehensively obtained. In general, the MBR of the pipeline is first determined by the stress at corner points of tension armor wires.

5.2 Structural response under complicated loads

When the pipeline is subjected to complicated multiaxial loads, the friction will be generated between the layers, which brings challenges to the evaluation of the overall mechanical responses of the pipeline. This section focuses on the mechanical behavior of tension armor wires under complicated loads, that is, considering the large interlayered pressure caused by the axial tension, internal and external pressure or torsion, and as the curvature increases from zero, the interlayered friction will force the steel wire on the surface of internal cylindrical core to go through three stages [17] of no-slip, stick–slip and full-slip, as shown in **Figure 10**. The theoretical model is based on the common assumptions: i) The lay angle of the wire remains constant during helical wire tension and bending, which yields a loxodromic curve on a torus surface; ii) The size of wire cross section is very small compared to the helical radius;

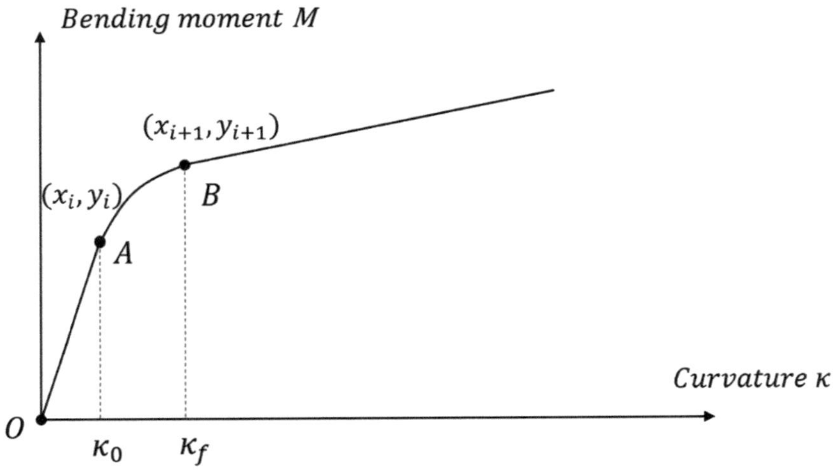

Figure 10.
Schematic diagram of nonlinear bending behavior of offshore flexible pipes.

iii) Small deformations occur in the linear elastic range of the material; iv) The helical wires are evenly distributed along the circumference and there is no mutual influence between neighboring wires.

5.2.1 No-slip stage (O-A)

When the curvature is very small, the helical wire remains firmly attached to the bent cylinder without any relative movement due to the interlayered static friction. Let R be the helical radius of the wire, α is the helical angle, θ represents the phase angle in the circumferential direction, which varies from zero to $\pi/2$. Considering the pipeline bent to a small curvature and the helical angle of the wire unchanged, the spatial three-dimensional coordinates $\kappa(x', y', z')$ of the deformed helical wire can be described parametrically as eq. (26) based on the differential geometry [18].

$$\begin{cases} x' = R\cos\theta \\ y' = \sin\left(\frac{R}{\tan\alpha}\theta\kappa\right)\frac{1}{\kappa} - R\sin\theta\sin\left(\frac{R}{\tan\alpha}\theta\kappa\right) \\ z' = \left(1 - \cos\left(\frac{R}{\tan\alpha}\theta\kappa\right)\right)\frac{1}{\kappa} + R\sin\theta\left(\frac{R}{\tan\alpha}\theta\kappa\right) \end{cases} \tag{26}$$

The corresponding axial strain along the bent helical strip is given by:

$$\varepsilon(\phi) = \frac{ds'}{ds} - 1 = \sqrt{\frac{dx'^2 + dy'^2 + dz'^2}{dx^2 + dy^2 + dz^2}} - 1 \tag{27}$$

where s represents the arch length along the helical wire. By using eq. (26) in eq. (27), and assuming small deflections, a linear expression for the axial strain can be obtained by eliminating second and higher order terms. The linearized axial strain is given by:

$$\varepsilon(\phi) = -R\cos^2\alpha\sin\theta\kappa \tag{28}$$

It can be seen that the maximum stress of the wire in the no-slip stage located at the intrados and extrados, and can be expressed as:

$$\sigma_{f-ns}^{max} = RE\cos^2\alpha\kappa \tag{29}$$

5.2.2 Stick–slip stage (A-B)

In the stick–slip stage, the axial force of the helical wire and the interlayered friction between are balanced with each other. However, as the curvature increases, the axial force on the wire section increases, and when the maximum static friction is exceeded, the wire begins to slide relative to internal components. Thus, the slipping condition can be written as [19]:

$$\frac{dN}{ds} > f \tag{30}$$

in which, N is the internal force along the wire direction, f means the friction force on the unit length of the wire, s and denotes the length along the wire axis. According to the eq. (28) and taking into account the geometric relationship $\theta = s \cdot \sin\alpha/R$, the critical curvature (Point A in **Figure 10**) corresponding to the wire starting to slip can be derived as:

$$\kappa_0 = \frac{f}{-EA\cos^2\alpha\sin\alpha\cos\theta} \tag{31}$$

It can be found that when θ equals to $k\pi$ ($k = 0, 1, 2, \ldots$), the critical curvature is minimized. That implies the helical wire at the neutral axis first beginning to slip as the curvature increasing. Assuming that the contact pressure on the inner and outer surfaces of the wire layer are separately q_i and q_o, and the corresponding static friction coefficients are μ_i and μ_o, the critical curvature of the wire can be further expressed as:

$$\kappa_0 = \frac{q_i\mu_i + q_o\mu_o}{-EA\cos^2\alpha\sin\alpha} \tag{32}$$

When the curvature is greater than the critical curvature, the slipping area of the wire rapidly extends from the neutral axis to both ends, during which one part of the wire is slipping while the other part still in the stick stage. According to the equilibrium relationship of the axial force and interlayered friction in the stick area, the initial slipping curvature at the phase angle θ of the wire can be obtained:

$$\kappa_f = \left(\frac{f}{-EA\cos^2\alpha\sin\alpha}\right)\frac{\theta}{\sin\theta} = \kappa_0\frac{\theta}{\sin\theta} \tag{33}$$

If $\theta = \pi/2$ in the above equation, the curvature when all parts of the wire enter the slipping state (Point B in **Figure 10**) can be described as:

$$\kappa_f = \frac{\pi}{2}\kappa_0 \tag{34}$$

From point A to point B, the wire completes the transition from the no-slip state to the full-slip state. As the axial stress of the wire is gradually released, the overall bending stiffness of the layer decreases.

5.2.3 Full-slip stage (B-)

Assuming that the friction along the wire direction dominates, the friction on the inner and outer surfaces of the rectangular wire produces the axial stress evenly distributed by the wire section. According to the eq. (33), the axial stress generated by the total friction F_0 of the slipping wire can be written as:

$$\sigma_{f-fs} = \frac{F_0}{A} = \left[\frac{R\theta}{A\sin\alpha}\right] f \tag{35}$$

where A is the cross-sectional area of the wire. It can be seen that the friction varies with the phase angle, but the stress at the four corner points of the wire at the same phase angle is the same. Therefore, the maximum axial stress of the wire generated by the total friction in the full-slip state is further obtained:

$$\sigma_{f-fs}^{max} = \frac{R\pi}{2A\sin\alpha}\left(q_i\mu_i + q_o\mu_o\right) \tag{36}$$

In summary, the theoretical model for calculating the maximum axial stress of the helical wire at each stage during the curvature increasing is established considering the interlayered friction. For conservative consideration in practical engineering, the above three stages are usually instead of two stages: one is the no-slip stage where the curvature varies from zero to κ_f., and the other is the full-slip section where the curvature is greater than κ_f

Additionally, due to the complicated loads, it is also necessary to consider the axial stress σ_T (as shown in Eq. (11)) of the wire caused by the tension and the axial stress σ_κ (as shown in Eq. (25)) caused by the helical "spring" itself during bending in the full-slip stage. Then the nonlinear stress of the helical wire under complicated loads with the consideration of the interlayered interaction can be expressed as Eq. (37) and shown as **Figure 11**.

$$\sigma = \begin{cases} \sigma_T + \sigma_{f-ns}, \kappa \leq \kappa_f \\ \sigma_T + \sigma_\kappa + \sigma_{f-fs}, \kappa > \kappa_f \end{cases} \tag{37}$$

In order to perform the fatigue evaluation [20], it is necessary to calculate the maximum alternating stress amplitude $K_c(\kappa)$ of offshore flexible risers as the curvature changes. Removing the average stress term σ_T and taking into account the maximum stress caused by the curvature and interlayered friction, the maximum alternating stress can be described as:

$$\Delta\sigma^{max} = \begin{cases} \sigma_{f-ns}^{max}, \kappa \leq \kappa_f \\ \sigma_\kappa^{max} + \sigma_{f-fs}^{max}, \kappa > \kappa_f \end{cases} \tag{38}$$

It can be seen that the alternating stress exhibits a nonlinear relation with the curvature. Therefore, $K_c(\kappa)$ is not a constant value and cannot be directly used in the existing commercial software to calculate the fatigue life. In this case, it is necessary to

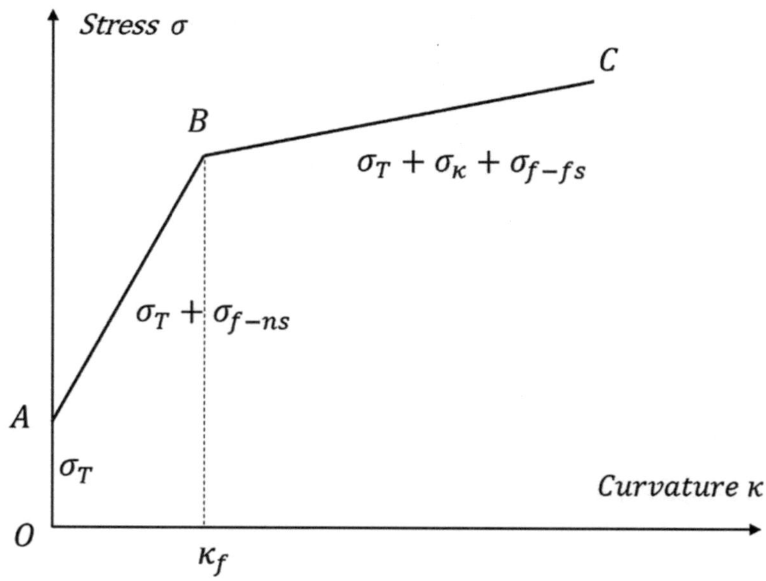

Figure 11.
Schematic diagram of nonlinear stress at the corner point of the wire cross-section.

carry out the secondary development in order to input the alternating stress with curvature, and then the fatigue life calculation can be performed by the damage accumulation using the Miner formula.

6. A case study

6.1 Description of the pipeline structure

In order to quantitatively illustrate the relation between interlayer action and the overall performance of offshore flexible pipelines, an 8-inch inner diameter flexible riser actually applied in a water depth of 1500 meters is taken in this section as an example, and the mechanical properties of pipelines are given in **Table 1**.

The pipeline structure is designed to have two tension armor layers, of which material elastic modulus is 206Gpa, the Poisson's ratio is 0.3 and the yield strength is 800Mpa. The rectangular cross-sectional dimensions of the steel wire in the two tension armor layers are the same (4 mm × 12 mm) and the helical angles are separately positive and negative 30 degrees. According to the configuration design, functional requirements and hydrodynamic analysis of the flexible riser, the design loads can be determined as the internal pressure of 20 Mpa, the external pressure of 15 Mpa, the tension of 170 tons, the MBR of 2.5 meters and the fatigue life of 25 years.

6.2 Analysis and discussion of structural responses

Using the theoretical models for calculating the interlayered interaction and structural responses proposed in this chapter, this section quantitively gives the

Layer	Thickness (mm)	Inside diameter (mm)	Outside diameter (mm)	Material
Carcass	8	200	216	AISI 316 L
Internal sheath	10	218	234	XLPE
Anti-wear 1	1	234	236	PA-11
Pressure armor	6	236	248	special carbon steel
Anti-wear 2	1	248	250	PA-11
Tension armor 1	4	250	258	special carbon steel
Anti-wear 3	1	258	260	PA-11
Tension armor 2	4	260	268	special carbon steel
Anti-wear 4	1	268	270	PA-11
External sheath 1	5	270	280	HDPE
External sheath 2	5	280	290	HDPE

Table 1.
Material and dimensional parameters of flexible pipelines.

interlayered pressure and the material utilization of the pipeline under axisymmetric and non-axisymmetric design loads, and the results are shown as in **Table 2**.

It can be clearly seen that the pipeline structure meets the design requirements, and still have a large safety margin under various independent loads. Those illustrate that the theoretical models for the structural responses proposed in this chapter could provide effective tools to design and evaluate the flexible pipelines in actual engineering. For the tension armor wires, achieving the tensile resistance of 170 tons requires the internal cylindrical component to provide a radial supporting pressure of nearly 6Mpa. Therefore, the internal pressure will be partially compensated by the pressure generated by axial tension, so that the material utilization of the tension armor wire is further reduced during actual operation of the pipeline.

Design load		Value	Interlayered pressure (MPa)	Stress (MPa)	Material utilization	Remarks
Axisymmetric	Internal pressure	20Mpa	19.67	403	50.4%	Only the pressure armor layer resisting
	External pressure	15Mpa	15.25	—	66.3%	Only the carcass layer resisting
	Tension	170 t	5.38Mpa (Inner surface)	371	46.4%	Distributed according to the stiffness of layers
			2.64Mpa (Outer surface)	372	46.5%	
Non-axisymmetric	MBR	2.5 m	Weak interlayered interaction	665	83.2%	In storage and installation
						The tenson armor layer dominate

Table 2.
Interlayered pressure and material utilization of the pipeline under different design loads.

Since the fatigue failure occurs first on the helical tension armor wire of offshore flexible riser under alternative loads, the nonlinear stress at the corner points of the wire with the curvature is calculated by using Eq. (38) under combining the axial tension and bending curvature, as shown by the red dot line in **Figure 12**. The figure also shows the linear stress (blue dot line) of the wire due to friction only in the no-slip state and the linear stress (black line) of the slipping wire without consideration of the interlayered friction. It can be clearly seen that the nonlinear wire stress considering interlayer friction is between the two linear stress distribution curves, so the corresponding distributions of the fatigue life should also have the similar tendency.

It should be noted that the interlayered contact and friction cause the stress of helical wires obvious nonlinear, so their accurate prediction is of great significance for the calculation of fatigue life. However, for the current analyzing theory of fatigue stress, there is a lack of accurate description of the wire behavior in the stick–slip stage [21], which has a decisive effect on the fatigue life when the riser is subjected to large axisymmetric loads and small bending curvature. In addition, after the wire enters the full-slip stage, it is worth further exploring whether the bending stress of the wire under axial tension can be simulated through the spring theory [22]. Moreover, despite the anti-wear layer, long-term repeated interlayered friction will still cause wear [23], which impact on the stress variation of the wire and the fatigue life cannot be ignored. All of those discussion items bring big challenges for evaluating the fatigue life of flexible pipes, which will promote the further development of the theoretical and experimental methods.

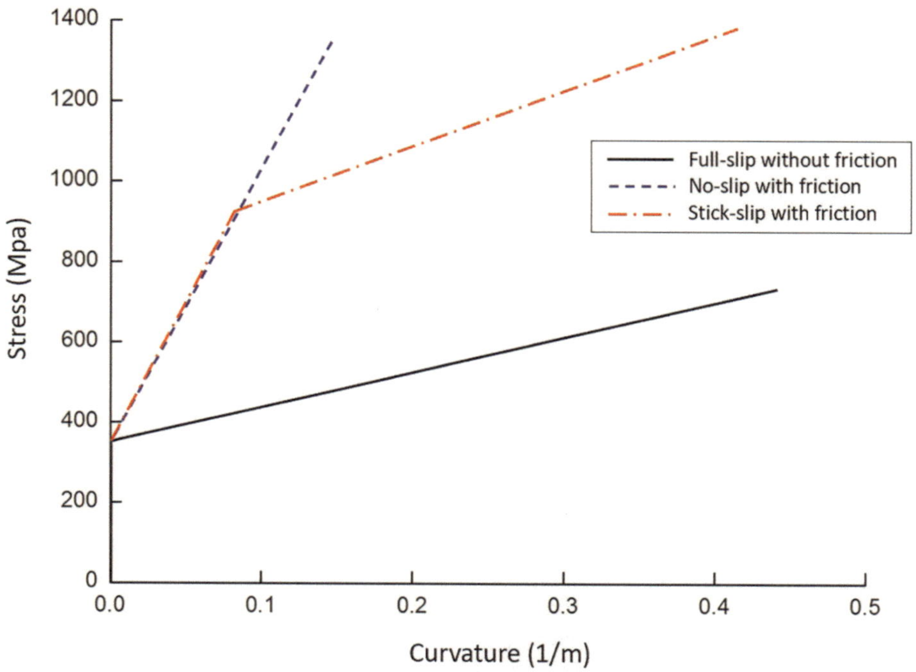

Figure 12.
Calculation results of nonlinear stress on the helical wire.

7. Conclusion

This chapter systematically introduces the mechanical principles of offshore flexible pipes, and gives the theoretical methods for evaluating the structural responses under axisymmetric loads, bending and combined complicated loads. It can be seen from a case study that although different strengthening layers in the pipeline are separately designed to resist against different loads, the overall resistance of the pipeline strongly depends on the interaction between the layers due to the use of metal helical armor wires, and then the unbonded pipe wall forms a whole. The contact pressure dominates the interlayered interaction under the axisymmetric loads such as axial tension, internal pressure, external pressure and torsion. And while the non-axisymmetric bending load is introduced, interlayered friction and relative slippage occur, resulting in the obvious stick–slip nonlinear behavior of pipeline components, which has an important impact on the subsequent fatigue life analysis.

Acknowledgements

This work was supported by the Program of NSFC (grant number 5170924); the Key Technologies Research and Development Program (grant number 2021YFC2801600); and the High-tech Ship Scientific Research Program (grant number CY05N20). Their support is gratefully appreciated.

Author details

Minggang Tang[1,2]

1 China Ship Science Research Center, Wuxi, China

2 Taihu Laboratory of DeepSea Technological Science, Wuxi, China

*Address all correspondence to: tangmg@cssrc.com.cn

References

[1] Zhang Y, Chen B, Qiu L, et al. Analytical tools optimize unbonded flexible pipes for Deepwater environments. Journal of Petroleum Technology. 2004;**56**(5):48-58

[2] International Organization for Standardization (ISO). Unbonded Flexible Pipe Systems for Subsea and Marine Applications. ISO-13628-2; (Switzerland) 2006

[3] De Sousa JRM, Lima ECP, Ellwanger GB, et al. Local mechanical behaviour of flexible pipes subjected to installation loads. In: Proceedings of 20th International Conference on Offshore Mechanics and Arctic Engineering. Rio de Janeiro, Brazil: ASME; 2001. pp. 219-227

[4] Tang M, Yang C, Yan J, et al. Validity and limitation of analytical models for the bending stress of a helical wire in unbonded flexible pipes. Applied Ocean Research. 2015;**50**:58-68

[5] Yue QJ, Lu QZ, Yan J, et al. Tension behavior prediction of flexible pipelines in shallow water. Ocean Engineering. 2013;**58**(15):201-207

[6] Knapp RH. Derivation of a new stiffness matrix for helically armoured cables considering tension and torsion. International Journal for Numerical Methods in Engineering. 1979;**14**(4): 515-529

[7] Gere JM, Goodno BJ. Mechanics of Materials. Toronto: Cengage Learning; 2009

[8] Zhu XK, Leis BN. Theoretical and numerical predictions of burst pressure of pipelines. Journal of Pressure Vessel Technology. 2007; **129**(4):644-652

[9] Timoshenko SP, Gere JM. Theory of Elastic Stability. Russia: St. Petersburg; 1961

[10] Zhang Y, Chen B, Qiu L, et al. State of the art analytical tools improve optimization of unbonded flexible pipes for Deepwater environments. In: Proceedings of the Offshore Technology Conference. Houston, Texas, USA: OTC; 2003

[11] Gay NA, de Arruda MC. A co MParative wet collapse buckling study for the carcass layer of flexible pipes. Journal of Offshore Mechanics and Arctic Engineering. 2012;**134**(3):031701

[12] Tang M, Lu QZ, Yan J, et al. Buckling collapse study for the carcass layer of flexible pipes using a strain energy equivalence method. Ocean Engineering. 2016;**2016**(111):209-217

[13] Tang MG, Yan J, Wang Y, et al. Tensile stiffness analysis on ocean dynamic power umbilical. China Ocean Engineering. 2014;**28**(2):259-270

[14] Roberto RJ, Pesce CP. A consistent analytical model to predict the structural behavior of flexible risers subjected to combined loads. Journal of Offshore Mechanics and Arctic Engineering. 2004;**126**(2):141-146

[15] Ramos R, Martins CA, Pesce CP, et al. Some further studies on the axial-torsional behavior of flexible risers. Journal of Offshore Mechanics and Arctic Engineering. 2014;**136**(1):011701

[16] Østergaard NH, Lyckegaard A, Andreasen JH. A method for prediction of the equilibrium state of a long and slender wire on a frictionless toroid applied for analysis of flexible pipe

structures. Engineering Structures. 2012; **34**:391-399

[17] Witz J, Tan Z. On the flexural structural behaviour of flexible pipes, umbilicals and marine cables. Marine Structures. 1992;**5**(2):229-249

[18] Knapp R. Helical wire stresses in bent cables. Journal of Offshore Mechanics and Arctic Engineering. 1988; **110**:55-61

[19] Kraincanic I, Kebadze E. Slip initiation and progression in helical armouring layers of unbonded flexible pipes and its effect on pipe bending behaviour. The Journal of Strain Analysis for Engineering Design. 2001;**36**(3): 265-275

[20] Sævik S. Theoretical and experimental studies of stresses in flexible pipes. Computers & Structures. 2011;**89**(23–24):2273-2291

[21] Tang M, Li SP, Zhang H, et al. Monitoring the slip of helical wires in a flexible riser under combined tension and bending. Ocean Engineering. 2022; **2022**(256):111512

[22] Zhang Y, Qiu L. Numerical model to simulate tensile wire behavior in unbonded flexible pipe during bending. In: Proceedings of the 26th International Conference on Offshore Mechanics and Arctic Engineering. San Diego, CA, United states: ASME; 2007. pp. 17-29

[23] Ye N, Sævik S, Zhou C. Anti-Wear tape and bending/fatigue of flexible risers. In: Proceedings of the 24th International Offshore and Polar Engineering Conference. Busan, Korea: ASME; 2014

Chapter 6

Terfenol-D Layer in a Functionally Graded Pipe Transporting Fluid for Free Vibration

Mukund A. Patil and Ravikiran Kadoli

Abstract

Knowledge of natural frequency of pipeline conveying fluid has relevance to designer to avoid failure of pipeline due to resonance. The damping characteristics of pipe material can be increased by using smart materials like magnetostrictive namely, TERFENOL-D. The objective of the present chapter is to investigate vibration and instability characteristics of functionally graded Terfenol-D layered fluid conveying pipe utilizing Terfenol-D layer as an actuator. First, the divergence of fluid conveying pipe is investigated without feedback control gain and thermal loading. Subsequently, the eigenvalue diagrams are studied to examine methodically the vibrational characteristics and possible flutter and bifurcation instabilities eventuate in different vibrational modes. Actuation of Terfenol-D layer shows improved stability condition of fluid conveying pipe with variation in feedback control gain and thermal loading. Differential quadrature and differential transform procedures are used to solve equation of motion of the problem derived based on Euler-Bernoulli beam theory. Finally, the effects of important parameters including the feedback control gain, thermal loading, inner radius of pipe and density of fluid on vibration behavior of fluid conveying pipe, are explored and presented in numerical results.

Keywords: control gain, isothermal load, flutter, bifurcation instability, differential quadrature and differential transform method

1. Introduction

Composite fluid-conveying pipes have become a practicable substitute to metallic pipes in several engineering applications such as oil and gas transport lines, hydraulic and pneumatic systems, thermal power plants, heat transfer equipment, petroleum and chemical process industries, underground refueling pipelines in airports, hospitals, medical devices, municipal sewage and drainage, corporation water supply and many more. Divergence and flutter instabilities are illustrious in fluid-conveying pipe due to fluid–structure interaction. One type of instability encountered in cantilever fluid-conveying pipes is called bifurcation, when the imaginary portion of the complex frequency disappears and the real portion splits into two branches. Fundamental concepts and early development in fluid structure interaction of fluid conveying pipes have been complied and studied by [1] systematically. A few more specialized topics are briefly discussed and well documented in Ref. [2–4]. Remarkable contributions in the area of fluid-conveying pipe vibrations also include the works of Chen [5].

In the meantime, performing a review on literature, it can be seen that a few studies have been carried out in the several field of vibrations such as in-depth nonlinear dynamics [6–10], vibration control [11–18], microtubes or nanotubes in microfluidic devices [19–22], and pipes using functionally graded materials [23–26].

The pseudo excitation method in conjunction with the complex mode superposition method was deduced to solve dynamic equation of Timoshenko pipeline conveying fluid [6]. The post-buckling and closed-form solutions to nonlinear frequency and response [8] of a FG fluid-conveying pipe have been investigated using analytical homotopy analysis method. Natural frequencies and critical flow velocities has been obtained for free vibration problem of pipes conveying fluid with several typical boundary conditions using DTM [11]. Dynamics and pull-in instability of pipes conveying fuid with nonlinear magnetic force have been investigated by [13], for clamped-clamped and clamped-free boundary conditions. The conclusion of investigation is that, location of magnets has a great impact on the static deflection and stability of the pipe. Wavelet based FEM has been used to examine the effect of internal surface damage [14] on free vibration behavior of fluid-conveying pipe. The natural frequencies of pipe conveying fluid has been determined by [15], using Muller's bisection method.

Failure due to filament wound with consideration of production process inconsistencies have been assessed by Rafiee et al. [16]. Vibration and instability response of magnetostrictive sandwich cantilever fluid-conveying micro-pipes is investigated utilizing smart magnetostrictive layers as actuators by [18].

Nonlinear vibration of a carbon nanotube conveying fluid with piezoelectric layer lying on Winkler-Pasternak foundation under the influence of thermal effect [21] and magnetic field [22] have been investigated using Galerkin and multiple scale method. The in-plane free vibration frequency of a zirconia-aluminum functionally graded curved pipe conveying fluid have been explored by the complex mode method [23]. The effect of axial variations of elastic modulus and density on dynamical behavior of an axially functionally graded cantilevered pipe conveying fluid has been analyzed by [24]. Dai et al. [25] studied the thermo-elastic vibration of axially functionally graded pipe conveying fluid considering temperature changes. Heshmati [26] studied the stability and vibration behaviors of functionally graded pipes conveying fluid considering the the effect of eccentricity imperfection induced by improper manufacturing processes. Xu Liang et al. [27] have used differential quadrature method (DQM) and the Laplace transform and its inverse, to analyze the dynamic behavior of a fluid-conveying pipe with different pipe boundary conditions. Huang Yi-min et al. [28] used the separation of variables method and the derived method from Ferrari's method to decouple the the natural frequency and the critical flow velocity equations of fluid-conveying pipe with both ends supported. Planar and spatial curved fluid-conveying pipe [29] have been investigated for their free vibration behavior with Timoshenko beam model and B-spline function used as the shape function in Galerkin method.

There are few investigations in the literature on fluid-conveying pipes containing Terfenol-D layers. Certainly, a study on the mechanical behavior of functionally graded Terfenol-D layered fluid conveying pipe will contribute to the understanding for future design engineers, hence an attempt on the vibration and stability of functionally graded Terfenol-D layered fluid conveying pipe. Inherent features of the Terfenol-D layer to regulate the vibration instabilities and critical flow velocity of a FGMT pipe are attempted numerically. Terfenol-D is a popular magnetostrictive material exhibiting force output for a corresponding magnetic field input and produces magnetic field for mechanical force as an input. Every term in Terfenol-D has a meaning (see **Figure 1**), for example, Ter means Terbium, Fe signifies chemical symbol for iron, Nol stands for Naval Ordnance Laboratory, and D stands for Dysprosium [30]. Terfenol-D has numerous distinguish characteristics, including a high electromechanical coupling coefficient (0.73), a high magnetostrictive strain (800–1600 ppm), a fast response, a high energy density, and a large output force. The total stiffness of the pipe is affected by actuation of the Terfenol-D layer due to the creation of tensile forces with a change in feedback control gain and temperature change in the fluid-conveying pipe. The governing equation of motion for FGMT fluid-conveying pipe is derived based on Euler-Bernoulli's theory. Differential quadrature and differential transform approaches are used to obtain the frequency of boundary value problem. Critical velocities of the FGMT pipe are also determined for various boundary conditions, feedback control gain, and thermal loading. Validation of frequencies and critical velocities is accomplished using accessible analytical relations.

2. Functionally graded fluid conveying pipe

Powder metallurgy is considered as manufacturing process for present functionally graded Terfenol-D layered fluid-conveying pipe. The functioanlly graded pipe is assumed to compose of aluminum (as metallic) and aluminum oxide (as ceramic). In between the graded composition of aluminum and aluminum oxide Terfenol-D layer is included. The material properties, volume fraction and expression for calculation of properties is given in [31]. **Figure 2** shows the layout of FGMT fluid-conveying pipe.

2.1 Derivation of governing equation

Considering the FGMT fluid-conveying pipe as an Euler-Bernoulli beam, the equation for the motion of the pipe can be derived using Hamilton's principle. The

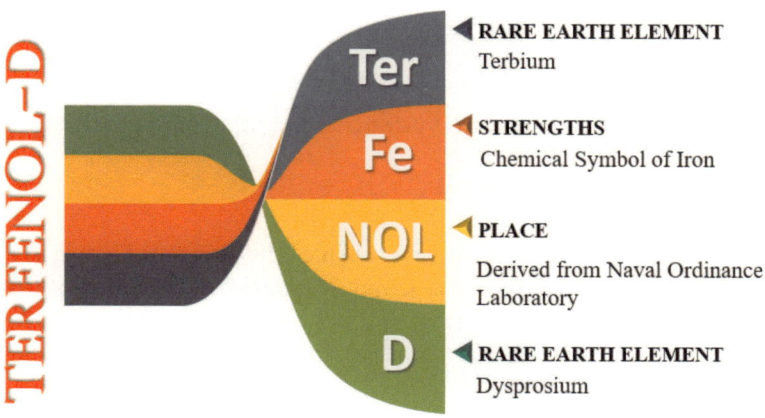

Figure 1.
Schematic for meaning of Terfenol-D.

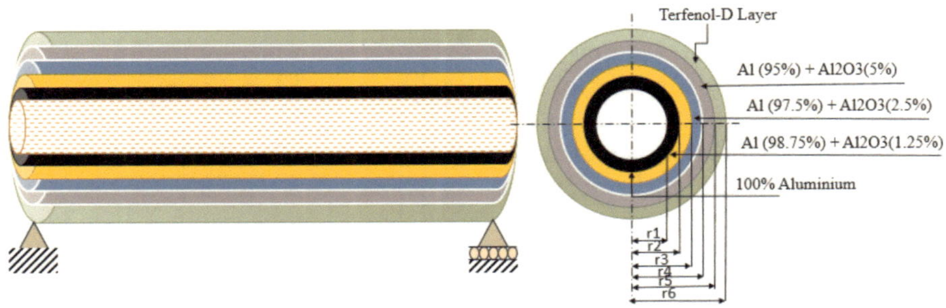

Figure 2.
Physical model of simply supported FGMT fluid-conveying pipe.

kinetic energy of the internal fluid is appended to the kinetic energy of the pipe to obtain total kinetic energy of FGMT pipe, and is described by the equation

$$J = J_p + J_f \tag{1}$$

Where, J_p and J_f signify the kinetic energy of the composite FGMT fluid-conveying pipe and the kinetic energy of the fluid flowing through the pipe, respectively. The elements of kinetic energy (J) as defined in Eq.(1) can be expressed as:

$$J_p = \frac{1}{2}\int_0^l m_p \left(\frac{\partial w}{\partial t}\right)^2 \mathrm{d}x \tag{2}$$

$$J_f = \frac{1}{2}\int_0^l m_f \left(\left(v\frac{\partial w}{\partial x} + \frac{\partial w}{\partial t}\right)^2 + v^2\right) \mathrm{d}x \tag{3}$$

Where, w symbolize for the displacement in the vertical direction, v symbolize for the fluid velocity, The flow of liquid, water, oil, and similar liquid flowing through the pipe are assumed to have a flat velocity profile at every section of the flow (i.e. popularly called as plug flow). m_p and m_f respectively denote the mass per unit length of the pipe and the internal fluid. The strain energy U of the fluid-conveying pipe can be defined as:

$$U = \frac{1}{2}\int_0^l E_p I_p \left(\frac{\partial^2 w}{\partial x^2}\right)^2 \mathrm{d}x \tag{4}$$

Where $E_p I_p$ is the flexural rigidity of the FGMT fluid-conveying pipe. Constitutive relation for a magnetostrictive beam type structure [32] could be written as:

$$\sigma_{xx}^T = C_{11}\epsilon_{xx} - e_{31}H_z \tag{5}$$

where σ_{xx}^T, ϵ_{xx} signifies axial stress and strain of the Terfenol-D layer. In addition, C_{11} and e_{31} are elastic stiffness coefficient and magnetostrictive constant, respectively. The subscript 31 indicates that, the magnetic field is applied in the 3(z) direction and mechanical response obtained in the 1(x) direction. The strength of the magnetic field H_z may now be stated as follows.

$$H_z = k_c C(t)\frac{\partial w}{\partial t} \tag{6}$$

Where, k_c, $C(t)$ and $\frac{\partial w}{\partial t}$ denotes the coil constant, feedback control gain and transverse displacement of fluid conveying pipe with respect to time, respectively. The strain energy of the Terfenol-D layer is given as:

$$U_T = \int_0^l \int_A \sigma_{xx}^T \epsilon_{xx} \mathrm{d}A dx \tag{7}$$

Also, the axial moment produced by Terfenol-D layer is,

$$M_{xx} = \int_A \sigma_{xx}^T z \mathrm{d}A \tag{8}$$

Applying the Hamilton's principle, one can write the functional of FGMT pipe as,

$$\int_{t_1}^{t_2} \delta(J - U - U_T)\mathrm{d}t + \int_{t_1}^{t_2} \delta W_{force} \mathrm{d}t = 0 \tag{9}$$

Where, J is the total kinetic energy of the system; U is the deformation energy of the system; W_{force} denotes the work of the non-conservative force. Therefore, the equation of motion for the free vibration of FGMT composite pipe conveying fluid can be written as:

$$\underbrace{E_pI_p\frac{\partial^4 w}{\partial x^4}}_{\text{Elastic}} + \underbrace{m_f v^2 \frac{\partial^2 w}{\partial x^2}}_{\text{Centrifugal}} + \underbrace{2m_f v \frac{\partial^2 w}{\partial x \partial t}}_{\text{Coriolis}} + \underbrace{\varepsilon \frac{\partial^2 w}{\partial x \partial t}}_{\text{Magnetostrictive Moment}} + \underbrace{\left(m_p + m_f\right)\frac{\partial^2 w}{\partial t^2}}_{\text{Inertia}} = 0 \tag{10}$$

The governing equation for FGMT fluid-conveying pipe with thermal loading making use of Ref. [33] can be obtained as:

$$E_pI_p\frac{\partial^4 w}{\partial x^4} + \left(m_f v^2 + A\gamma(\Delta T)\right)\frac{\partial^2 w}{\partial x^2} + 2m_f v \frac{\partial^2 w}{\partial x \partial t} + \varepsilon \frac{\partial^2 w}{\partial x \partial t} + \left(m_p + m_f\right)\frac{\partial^2 w}{\partial t^2} = 0 \tag{11}$$

Where,

$$m_p = \sum_{j=1}^{n} \pi\rho_j \left(r_{j+1}^2 - r_j^2\right) \tag{12}$$

$$E_pI_p = A_{11}r^3 - D_{11}r \qquad r = \frac{d_o + d_i}{4} \tag{13}$$

$$A_{11} = \sum_{j=1}^{n} Q_{11}\left(r_{j+1} - r_j\right) \tag{14}$$

$$D_{11} = \frac{1}{3}\sum_{j=1}^{n} Q_{11}\left(r_{j+1}^3 - r_j^3\right) \tag{15}$$

$$\varepsilon = e_{31}k_c C(t)\left(r_{j+1}^2 - r_j^2\right) \tag{16}$$

$$\gamma(\Delta T) = E\alpha\Delta T \tag{17}$$

Where, α indicates the thermal expansion coefficient of the fluid conveying pipe material, ΔT is the temperature change in the layers, E is the Young's modulus of the fluid conveying pipe and $\gamma(\Delta T)$ symbolize the linear elastic stress–temperature coefficient.

3. Transformation of PDE into a sets of ODEs

Authors used the differential quadrature method to solve the free vibration equation of FGMT fluid-conveying pipe as given in Eq. (9). Here, the Eq. (9) is transformed into sets of ordinary differential equations. The standard eigenvalue form [34, 35] of the Eq. (9) can be obtained by assuming:

$$w_0 = W_0 e^{\Lambda t} \tag{18}$$

W_0 is the mode shape of transverse motion and Λ is the frequency of the FGMT fluid-conveying pipe. Substitute the Eq. (12) in Eq. (9), accordingly Eq. (9) re-reads as follows:

$$E_p I_p \frac{\partial^4}{\partial x^4}\left(W_0 e^{\Lambda t}\right) + m_f v^2 \frac{\partial^2}{\partial x^2}\left(W_0 e^{\Lambda t}\right) + 2 m_f v \frac{\partial}{\partial x}\left(\frac{\partial}{\partial t} W_0 e^{\Lambda t}\right) + \epsilon \frac{\partial}{\partial x}\left(\frac{\partial}{\partial t} W_0 e^{\Lambda t}\right) + (m_p + m_f)\frac{\partial^2}{\partial t^2}\left(W_0 e^{\Lambda t}\right) = 0 \tag{19}$$

$$\left(E_p I_p \frac{d^4 W_0}{dx^4} e^{\Lambda t}\right) + \left(m_f v^2 \frac{d^2 W_0}{dx^2} e^{\Lambda t}\right) + \left(2 m_f v \frac{d W_0}{dx} e^{\Lambda t}\right)\Lambda + \left(\epsilon \frac{d W_0}{dx} e^{\Lambda t}\right)\Lambda + \left((m_p + m_f) W_0 e^{\Lambda t}\right)\Lambda^2 = 0 \tag{20}$$

$$\left(E_p I_p \frac{d^4 W_0}{dx^4}\right) + \left(m_f v^2 \frac{d^2 W_0}{dx^2}\right) + \left((2 m_f v^2 + \epsilon)\frac{d W_0}{dx}\right)\Lambda + \left((m_p + m_f) W_0\right)\Lambda^2 = 0 \tag{21}$$

Now, substitute the analog form of differential quadrature for respective derivative (first, second, third and fourth) such as:

$$\frac{d^4 W_0}{dx^4} = \sum_{j=1}^{N} A_{ij}^{(4)} W_j, \qquad \frac{d^2 W_0}{dx^2} = \sum_{j=1}^{N} A_{ij}^{(2)} W_j, \qquad \frac{d W_0}{dx} = \sum_{j=1}^{N} A_{ij}^{(1)} W_j \tag{22}$$

Now, Eq.15 becomes,

$$E_p I_p \sum_{j=1}^{N} A_{ij}^{(4)} W_j + m_f v^2 \sum_{j=1}^{N} A_{ij}^{(2)} W_j + \left((2 m_f v^2 + \varepsilon)\sum_{j=1}^{N} A_{ij}^{(1)} W_j\right)\Lambda + \left((m_p + m_f) W_i\right)\Lambda^2 = 0 \tag{23}$$

Now separate the terms associated with Λ and Λ^2 to prepare the damping and mass matrices, respectively as shown in Eq. 18.

$$\{-[M]\Lambda^2\}\{d\} + \{[\Gamma]\Lambda\}\{d\} + [K]\{d\} = 0 \tag{24}$$

Where,

$$[\Gamma] = [C_{dd}] - [C_{db}][S_{bb}]^{-1}[S_{bd}] \tag{25}$$

$$[K] = [S_{dd}] - [S_{db}][S_{bb}]^{-1}[S_{bd}] \tag{26}$$

Where, C_{dd} and C_{db} are the damping sub matrices which includes the domain-domain and domain-boundary elements of damping. Similarly, S_{bb}, S_{bd}, S_{db} and S_{dd} are the stiffness sub matrices which includes the boundary-boundary, boundary-domain, domain-boundary and domain-domain elements, respectively. The standard form of eigenvalue can be obtained from Eq. (18) as:

$$\left\{\begin{bmatrix} 0 & I \\ \Gamma & K \end{bmatrix} - \begin{bmatrix} I & 0 \\ 0 & M \end{bmatrix}\Lambda\right\}\left\{\begin{matrix} d \\ \Lambda d \end{matrix}\right\} = 0 \tag{27}$$

Where I, $[K]$, $[\Gamma]$ and $[M]$ denote the identity, structural stiffness, damping and mass matrix, respectively. One can obtain the two sets of eigenvalues. The eigenvalue obtained can be written as $\Lambda = -\alpha \pm i\omega_d$.

4. Application of differential transform method to FGMT fluid-conveying pipe

Differential transform technique (DTM) may be used to solve integral equations, ordinary partial differential equations, and differential equation systems. Using this approach, a polynomial solution to differential equations may be derived analytically. For large orders, the Taylor series approach is computationally time-consuming. This method is appropriate for linear and nonlinear ODEs since it does not need linearization, discretization, or perturbation. It is also possible to significantly reduce the amount of computing labour required while still precisely delivering the series solution and rapidly converging. The DTM has several disadvantages, though. Using the DTM, a truncated series solution may be obtained. This truncated solution does not display the actual behavior of the problem, but in the vast majority of situations it offers a good approximation of the actual solution in a relatively limited area. Solutions are expressed as convergent series with components that may be readily computed using the differential transform technique. The linear equation of motion for free vibration of FGMT fluid-conveying pipe is given by,

$$\left(E_pI_p\frac{d^4W_0}{dx^4}\right) + \left(m_fv^2\frac{d^2W_0}{dx^2}\right) + \left(\left(2m_fv^2 + \varepsilon\right)\frac{dW_0}{dx}\right)\Lambda + \left(\left(m_p + m_f\right)W_0\right)\Lambda^2 = 0 \tag{28}$$

The differential transformation form of Eq. (22) can be written as

$$\begin{aligned} E_pI_p((i+1)(i+2)(i+3)(i+4)W(i+4)) + m_fv^2((i+1)(i+2)W(i+2)) \\ +(2m_fv^2 + \varepsilon)((i+1)W(i+1)) + (m_p + m_f)W(i) = 0 \end{aligned} \tag{29}$$

$x = 0$		$x = 1$	
Original Form	**DTM Form**	**Original Form**	**DTM Form**
$w(0)=0$	$W(0)=0$	$w(1)=0$	$\sum_{i=0}^{N} W(i)=0$
$\frac{dw}{dx}(0)=0$	$W(1)=0$	$\frac{dw}{dx}(1)=0$	$\sum_{i=0}^{N} iW(i)=0$
$\frac{d^2w}{dx^2}(0)=0$	$W(2)=0$	$\frac{d^2w}{dx^2}(1)=0$	$\sum_{i=0}^{N} i(i-1)W(i)=0$
$\frac{d^3w}{dx^3}(0)=0$	$W(3)=0$	$\frac{d^3w}{dx^3}(1)=0$	$\sum_{i=0}^{N} i(i-1)(i-2)W(i)=0$

Table 1.
Transformed form of boundary condition for differential transform method.

Rearranging Eq. (23), one will get a simple recurrence relation as:

$$W(i+4) = -\frac{\left(m_f v^2 (i+1)(i+2)W(i+2) + \left(2m_f v^2 + \varepsilon\right)(i+1)W(i+1) + \left(m_p + m_f\right)W(i)\right)}{E_p I_p (i+1)(i+2)(i+3)(i+4)} \tag{30}$$

Similarly, analogous form of original boundary conditions for the differential transformation can be done using **Table 1**, where $x = 0$ and $x = 1$ represents the boundary points. It can be seen that $W(i)$, $(i = 4{,}5, \ldots ,N)$ is a linear function of $W(2)$ and $W(3)$. Thus, $W(2)$ and $W(3)$ are considered as unknown parameters and taken as $W(2) = b_1$, $W(3) = b_2$ for clamped-clamped boundary conditions. With Eq. (23), $W(i)$ can be calculated via an iterative procedure. Substituting $W(i)$ into boundary conditions at other end of FGMT pipe, the two equations (Substituting all $W(i)$ terms into boundary condition expressions) can be written as matrix form,

$$\begin{bmatrix} R_{11} & R_{12} \\ R_{21} & R_{22} \end{bmatrix} \begin{bmatrix} b_1 \\ b_2 \end{bmatrix} = 0 \tag{31}$$

Where R_{ij} are associated with the eigenvalues ω, b_1 and b_2 are the constants and other parameters of the FGMT pipe system, corresponding to N. To obtain a non-trivial solution of Eq. (25), it is required that the determinant of the coefficient matrix vanishes, namely

$$\begin{vmatrix} R_{11} & R_{12} \\ R_{21} & R_{22} \end{vmatrix} = 0 \tag{32}$$

Therefore, the eigenvalues ω can be computed numerically from Eq. (26). Generally, ω is a complex number.

5. Results and discussion

In the following section, the numerical results are proposed to investigate the free vibration behavior of FGMT fluid-conveying pipe subjected to control gain and thermal loading. Since there is no published research on the subject of free vibration of FGMT fluid-conveying pipes in the open literature, a differential quadrature and differential transform approach is used to conduct a condensed analysis of the current study. The imaginary component ($\mathfrak{Im}$) of the complex frequency $[\Omega = \mathfrak{Re}(\Omega) \pm \mathfrak{Im}(\Omega)]$ denotes the energy stored in either mass or strain energy in the fluid conveying pipe. The accumulated strain energy is linked to the failure behavior of the fluid conveying pipe. Furthermore, the real element ($\mathfrak{Re}$) of the complex frequency represents damping and the energy that will be transformed to heat or other energy by friction or other molecular actions.

5.1 Validation of present study

The current MATLAB code for the differential quadrature and transform technique is validated using Ref. [28], as shown in **Table 2**. The validation for FGMT fluid conveying pipe is also given by the author in Ref. [36]. Furthermore, the solution obtained using the differential quadrature approach corresponds well with the solution acquired using the differential transform method.

It has been identified that, the differential transform method requires the 58 number of terms to get the converged solution whereas 19 grid points used to obtain the convergence solutions shown. The natural frequencies of pipes conveying fluid depend on the fluid velocity v. The physical parameters of FGMT fluid-conveying pipe are calculated as: $m_p = 1.0670$ kg − m, $m_f = 0.23562$ kg − m, $E_pI_p = 5.1620$ N − m^2, $L = 1$ m. In order to calculate these physical parameters, authors have used Eq. 2.1. MATLAB software is used to create a package that performed the foregoing computations. The correctness of the results are shown by the comparison of the results of differential transform method in **Table 3** under different boundary conditions for $v = 0.5$ m/s. The number of grid points was modified from 7 to 19 to reach the converged solution. From the **Table 3**, it can be concluded that the imaginary component of the damped frequency calculated using DQM and DTM coincides rather well.

One of the key concerns for fluid conveyance pipes to be of significant importance is stability. The natural frequencies decrease with higher flow rates for pipelines with supported ends. The system destabilizes by diverging (buckling) when the natural frequencies fall to zero, and the resulting flow velocity is known as the critical flow velocity. In the case of $v \neq 0$, **Figures 3–10** represent the natural frequencies of fluid-conveying FGMT pipe with different boundary conditions. The first three natural

Velocity (m/s)	Method	Mode	
		λ_1	λ_2
	DQM	15.7213	31.4327
v=0	DTM	15.2765	31.3274
	Ref. [28]	15.71	31.42
	DQM	13.9650	30.8266
v=10	DTM	13.5669	30.6656
	Ref. [28]	13.97	30.83

Table 2.
Validation of simply-supported natural frequencies (rad/sec) of fluid conveying pipe (Parameters used: EI = 100 Nm^2, m$_f$ = 2 kg/m, m$_p$ = 2 kg/m *and L = 1 m).*

Boundary	Nodes	Mode			
		$Im(\Lambda_1)$	$Im(\Lambda_2)$	$Im(\Lambda_3)$	$Im(\Lambda_4)$
	7	18.4437	58.7726	111.9680	—
	11	18.4530	76.9717	166.6046	255.9565
S-S	15	18.4529	77.4675	175.6156	331.6998
	17	18.4529	77.4688	175.7221	313.2417
	19	18.4529	77.4688	175.7221	313.2417
DTM		18.4765	77.4445	175.6849	313.2167
	7	44.6652	94.0809	149.5829	—
	11	43.8719	120.7255	224.0424	324.0114
C-C	15	43.8700	121.9308	239.5009	435.8037
	17	43.8700	121.9365	239.7773	396.9056
	19	43.8700	121.9365	239.7780	396.9056
DTM		43.9087	122.1424	240.0518	397.2302
	7	29.5136	75.3884	133.9371	—
	11	29.8002	97.5593	196.8662	272.6251
S-C	15	29.8007	98.5000	206.4495	391.9600
	17	29.8006	98.5037	206.5287	353.8381
	19	29.8006	98.5037	206.5287	353.8381
DTM		29.8321	98.6121	206.6739	354.0290
	7	5.1817	41.1890	112.4169	—
	11	7.1259	43.1741	128.5438	207.9994
C-F	15	7.1260	43.1894	122.0607	241.8281
	17	7.1260	43.1894	121.9887	239.7690
	19	7.1260	43.1894	121.9887	239.7690
DTM		7.1487	43.5897	122.3256	240.9832

Table 3.
Convergence of imaginary component of damped frequency for different boundary conditions when $v = 5$ m/s.

frequencies of the C-F fluid-conveying FGMT pipe with $0 \leq v \leq 50$ are depicted in **Figures 3** and **4**. The critical velocity of the pipe is $v = 42$ m/s, and the third mode appears flutter instability. The findings of the differential quadrature method were utilized to plot the results presented in the **Figures 3–10**.

The first four natural frequencies of the simply supported fluid-conveying FGMT pipe with $0 \leq v \leq 50$ are plotted in **Figures 5** and **6**. The first mode appears divergence instability when the critical velocity of the FGMT pipe is $v = 15$ m/s, and Paidoussis coupled mode flutter instability appears when the critical velocity is $v = 31$. Real component ($\mathfrak{Re}$) of the complex frequency is almost zero during the first mode divergence instability. By increasing the flow velocity 30 m/s, the imaginary part of combination of first and second modes becomes zero, while the real part is non-zero, and the non-zero frequency and damping of first and second mode at the same values are coupled, then the system will be unstable again. This sort of instability, caused by

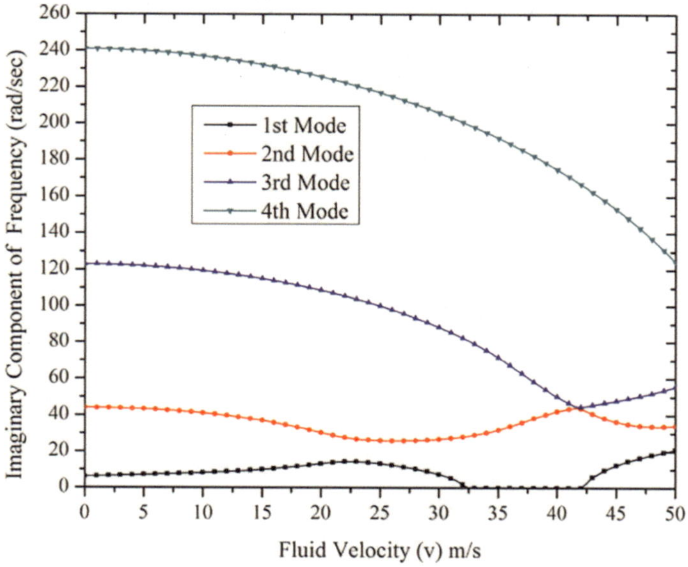

Figure 3.
Effect of fluid velocity v on imaginary component of clamped-free damped frequency.

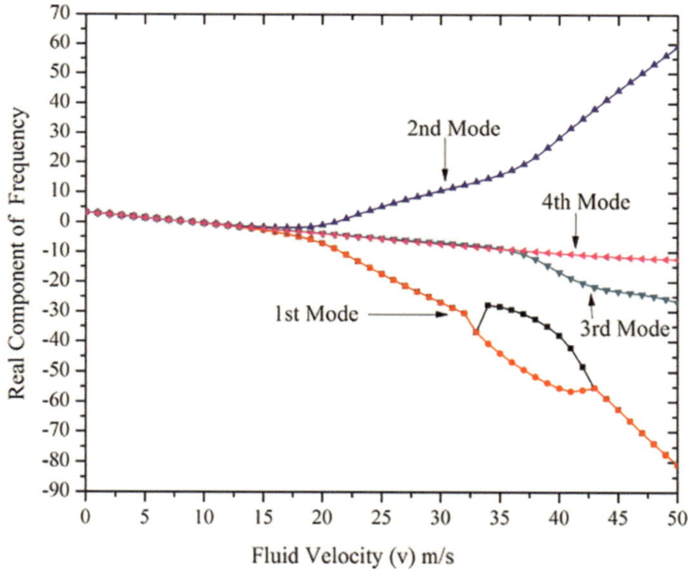

Figure 4.
Effect of fluid velocity v on real compoent of clamped-free damped frequency.

the interaction of two modes, is known as flutter instability, and its amplitude develops exponentially as a function of time.

Figures 7 and **8** shows the first four natural frequencies of the C-C fluid-conveying pipe with $0 \leq v \leq 50$. The critical velocity of the FGMT pipe is $v = 30$ m/s and 43, and corresponds to divergence instability in the first mode and couple-mode flutter instability. Bifurcation critical flow velocity is the term used to describe the flow velocity at which the bifurcation instability occurs. It should be noted that the system enters an

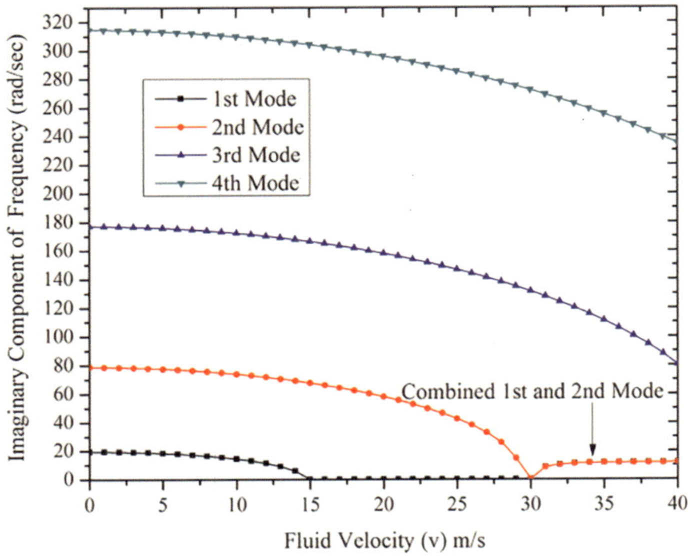

Figure 5.
Effect of fluid velocity v on imaginary component of simply supported damped frequency.

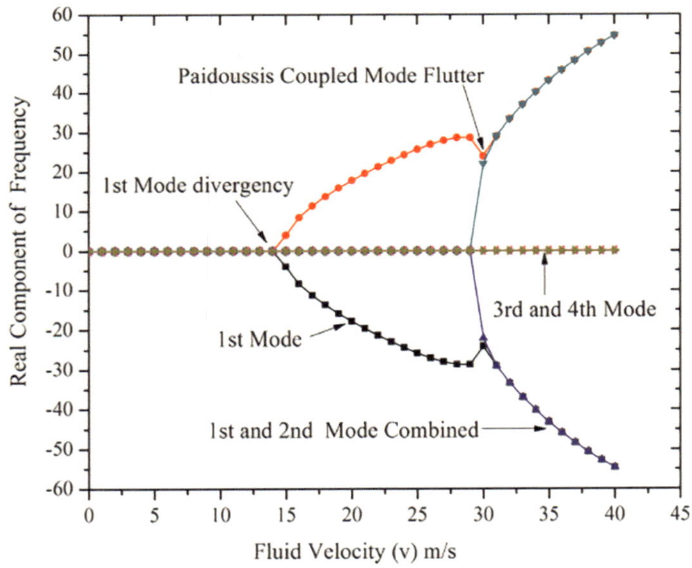

Figure 6.
Effect of fluid velocity v on real component of simply supported damped frequency.

over-damping mode, which prevents the FGMT pipe from vibrating, when the working fluid velocity surpasses its critical value.

Figures 9 and **10** presents the first four natural frequency of the S-C fluid-conveying FGMT pipe with $0 \leq v \leq 50$. It is obvious that the first mode appears divergence instability when fluid velocity $v = 22$ m/s, and coupled-mode flutter instability appears when fluid velocity reaches to $v = 37$ m/s. The specific critical velocities under different boundary conditions are listed in **Table 4**. The critical velocity for the

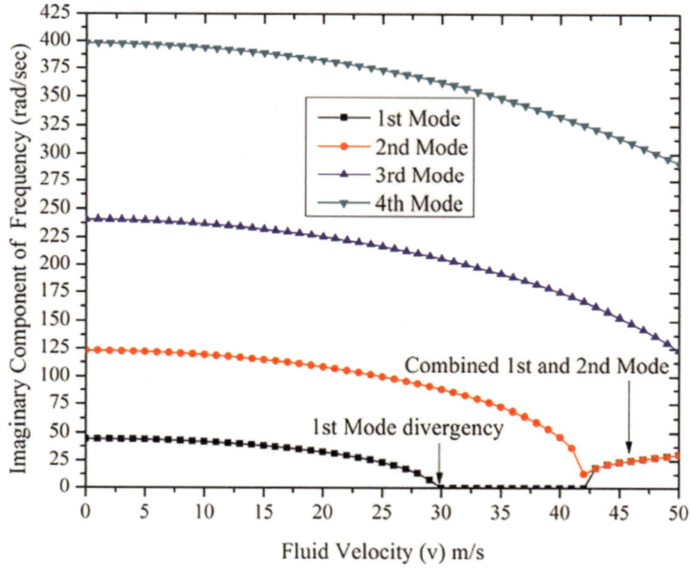

Figure 7.
Effect of fluid velocity v on imaginary component of clamped-clamped damped frequency.

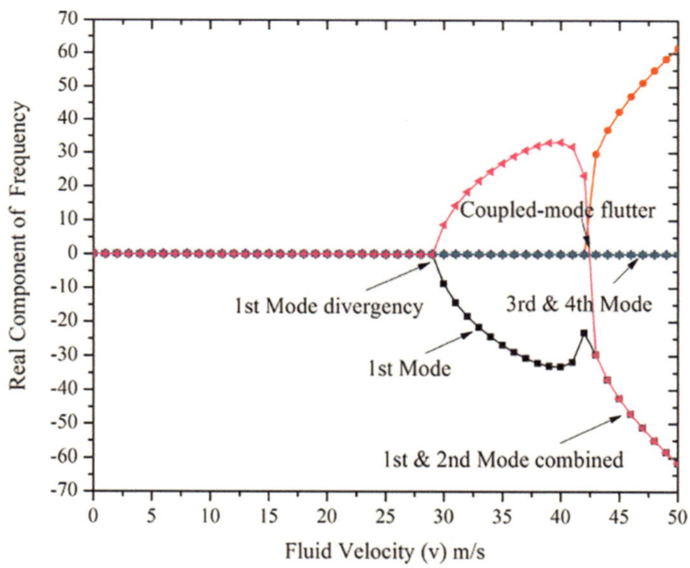

Figure 8.
Effect of fluid velocity v on real component of clamped-clamped damped frequency.

simply supported-simply supported and clamped-clamped boundary conditions are validated using Navier solution given by [37].

The relationships between the imaginary component of frequency of the FGMT pipe and the fluid density for different boundary conditions are plotted in **Figure 11**. Because the inertial and Coriolis forces were stronger with increasing fluid density, it was more simpler for the pipe to lose its stability. This led to a lower natural frequency. The changes of imaginary component of frequency with inner radius of the

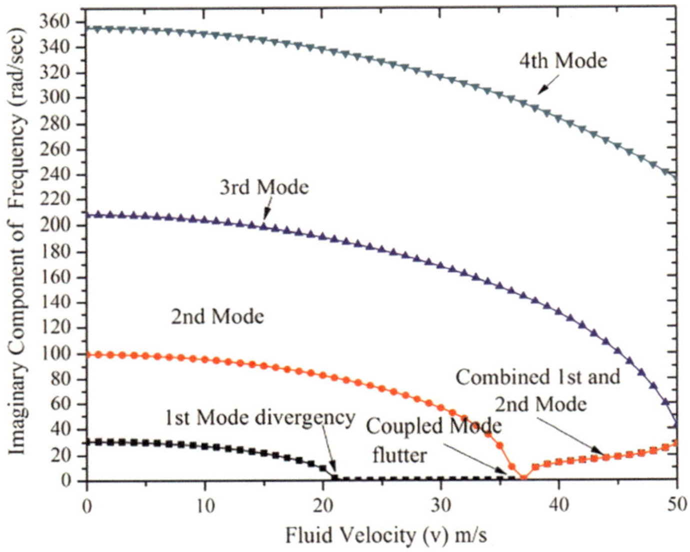

Figure 9.
Effect of fluid velocity v on imaginary component of simply supported-clamped damped frequency.

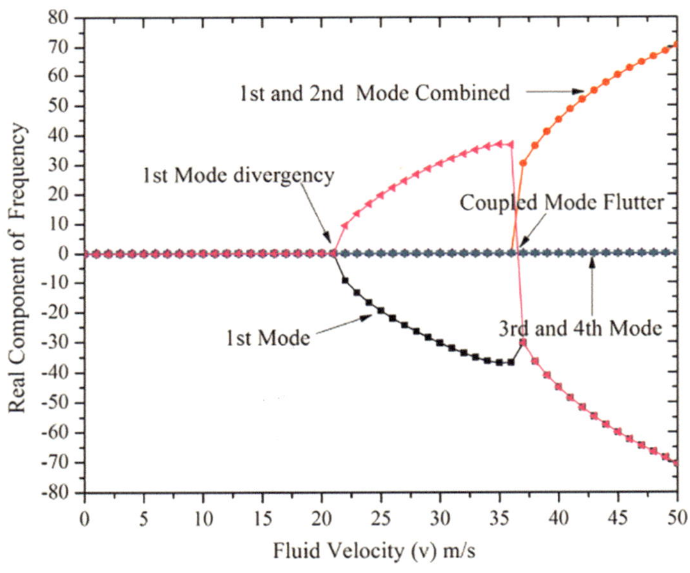

Figure 10.
Effect of fluid velocity v on real component of simply supported-clamped damped frequency.

FGMT pipe for different boundary conditions are shown in **Figure 12**. For very small values of the inner radius, an increase in the inner radius has a considerable impact on frequency; nevertheless, when the inner radius value is near to the outer radius, the frequency increases. In the boundary conditions clamped-clamped, simply supported-simply supported, and simply supported-clamped, the imaginary component of frequency drops as the feedback control gain rises. Imaginary component of the eigenvalue for a clamped-free frequency becomes zero for 3000 feedback control gain, $r = 0.005$ m and $v = 5$ m/s shown in **Figure 13**.

BC	Mode	Velocity (v)	Instability Form
S-S	1st Mode	15	Divergence
	Navier Solution [37]	15	—
	2nd Mode	30	Divergence
	1st & 2nd Combined	31	Paidoussis coupled mode flutter
C-C	1st Mode	30	Divergence
	Navier Solution [37]	30	—
	1st & 2nd Combined	43	Coupled mode flutter
C-F	3rd Mode	42	Flutter
S-C	1st Mode	22	Divergence
	1st & 2nd Combined	37	Coupled mode flutter

Table 4.
Critical velocities for FGMT pipe with different boundary conditions.

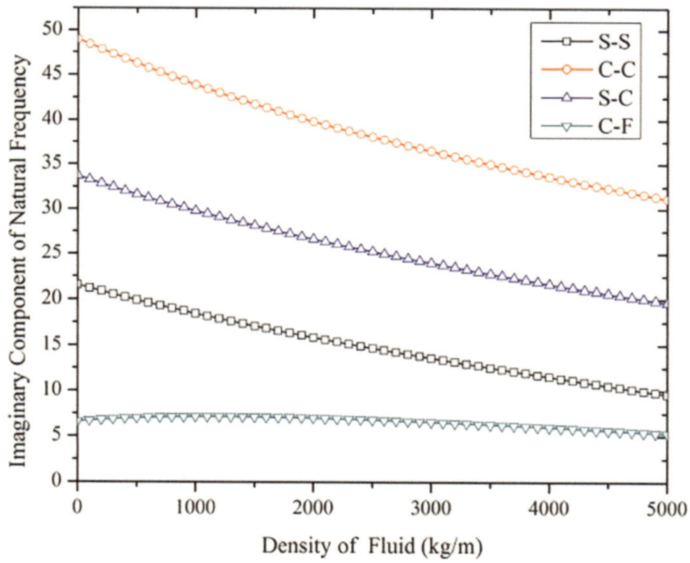

Figure 11.
Variation in fundamental natural frequency of FGMT pipe with changes in fluid density.

It is worth pointing out that the important aspect of present research work is maneuvering the use of Terfenol-D layers attached on the top FGMT fluid-conveying pipe to control the critical flow velocity and also improve the stability region. When Terfenol-D layer actuates tensile forces are generated in FGMT fluid-conveying pipe which affects the stiffness of fluid-conveying pipe. In order to evaluate this objective, **Figure 14** shows the real part ($\Re e$) of clamped-free first mode frequency with flow velocity for 0, 1000 and 1500 feedback control gain. It is observed that, 30, 28 and 9 m/s are the critical flutter velocity for 0, 1000 and 1500 feedback control gain, respectively. Therefore, one can make fluid-conveying pipe more stable by varying the feedback control gain. **Figure 15** shows the variation of analytical nonlinear

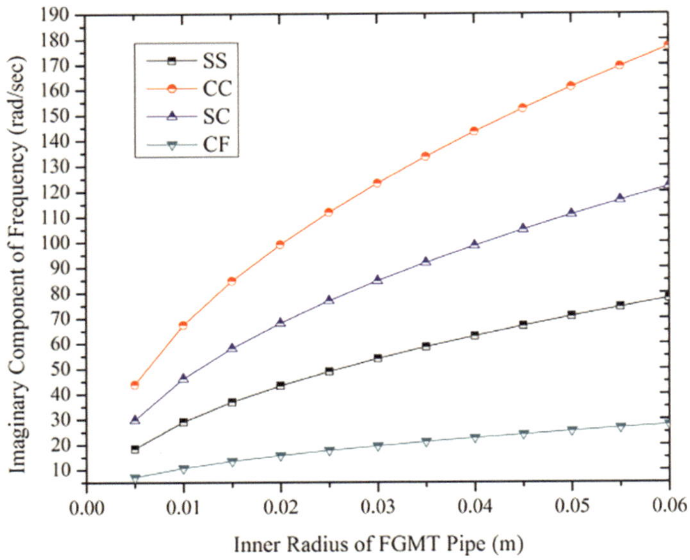

Figure 12.
Variation of fundamental frequency with changes in inner radius of FGMT pipe for different boundary conditions.

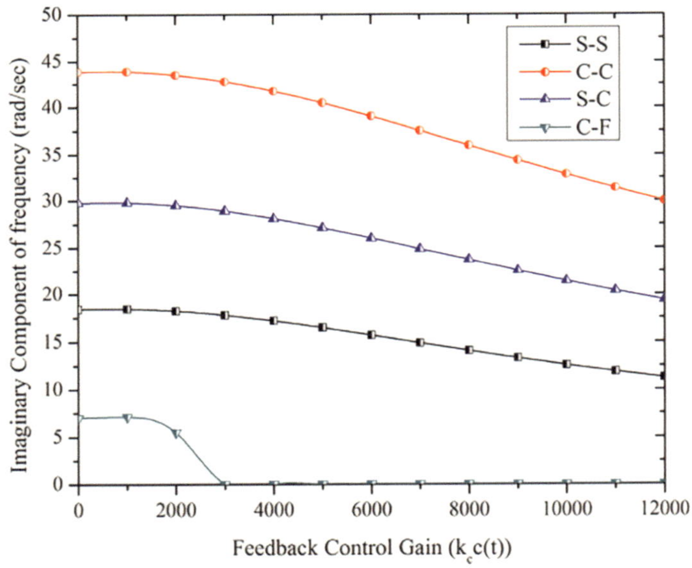

Figure 13.
Variation of imaginary component of the frequency with changes in feedback control gain at $r = 0.005$ m and $v = 5$ m/s.

frequency of FGMT fluid conveying pipe calculated based on relations published by [38] for simply supported boundary condition. It has been shown that when fluid velocity rises, the nonlinear frequency falls.

Figure 16 depicts the coupled effect of feedback control gain along with thermal loading. It is inferred that, there is decreasing effect of critical flow velocity as thermal loading increases. The reduction in overall stiffness of pipe is the reason for instability of FGMT pipe at lower flow velocity with thermal loading. Therefore, critical flow

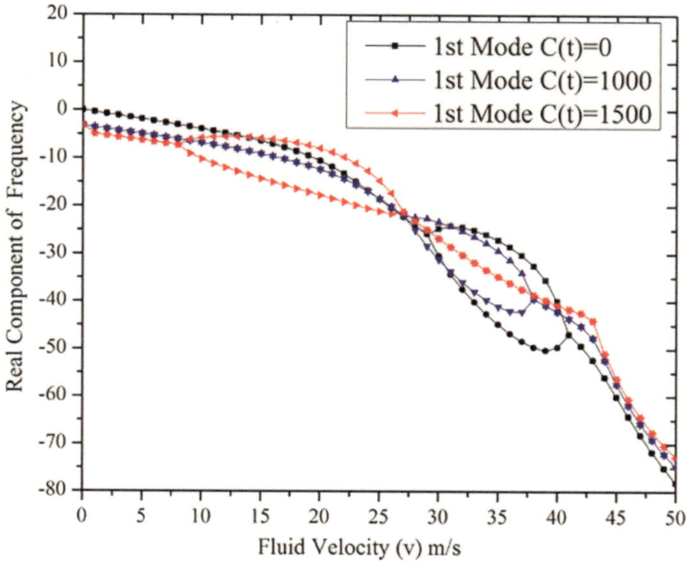

Figure 14.
Variation of clamped-free fundamental frequency with changes in control gain and fluid velocity.

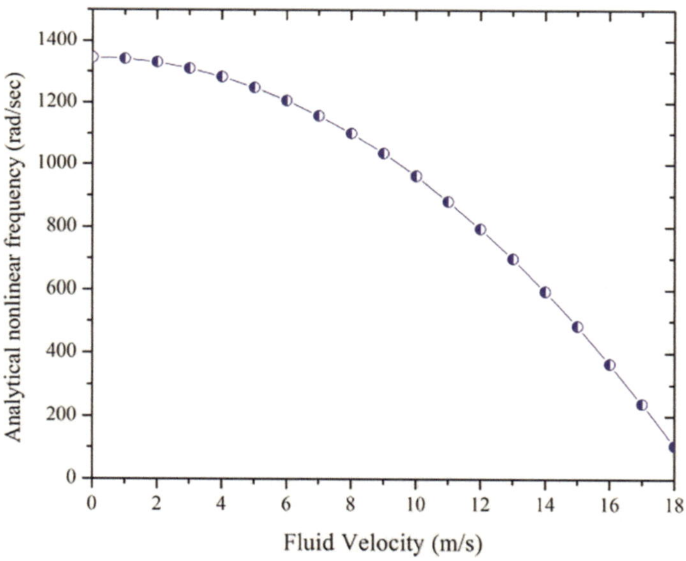

Figure 15.
Variation of nonlinear simply supported frequency with changes in fluid velocity.

velocity condition under thermal loading can be amplified through imposing higher feedback control gain. The control gain varies between 0 and 2000 as the temperature of the fluid conveying pipe changes. It is inferred that, with a zero control gain and 1°C and 0°C, the instability state of the fluid conveying pipe reduces from a fluid velocity from 27 to 25.2 m/s. Additionally, with a control gain of 1000, the fluid conveying pipe's unstable condition decreases from 29 to 24 m/s. Similar to this, with the control gain of 2000, the fluid conveying pipe's unstable condition decreases from 30 to 22 m/s.

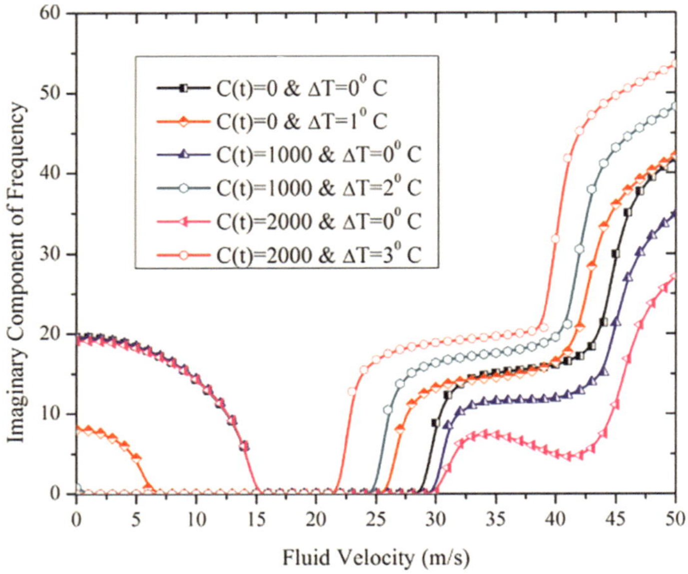

Figure 16.
Variation of S-S fundamental frequency with changes in control gain, thermal loading and fluid velocity.

6. Concluding remarks

In this chapter, the differential quadrature and differential transform method is applied to analyze the free vibration of FGMT pipes conveying fluid with different boundary conditions. Boundary value problem of FGMT fluid-conveying pipe is solved straightforwardly using DQM and DTM. Close agreement is established for critical velocity and frequencies results generated by DQM, DTM with those of Navier and Galerkin solution. Eigenvalue diagrams are detailed enough to shows the illustration about the effects of feedback control gain, density of fluid, inner radius of pipe and thermal loading on the vibrational and instability characteristics. To attenuate the amplitude of vibration or displacement, inherent damping property of the material cannot be sufficient. To dampen out large amplitude vibration during resonance, special techniques have been explored, like using sandwich pipes namely, viscoelastic layer placed between two layers of the parent pipe material. This approach is called passive damping. Viscoelastic materials like, natural rubber, and synthetic rubber like nitrile butadine rubber and styrene butadine rubber, silicone rubber can be proposed. Sophisticated technique is the active vibration. This method involves use of materials like, piezoelectric, magnetostrictive, magnetorehology, electrostricitve and shape memory alloys. Magnetostrictive material presented in this chapter works on the ability of the material to respond mechanically to the presence of magnetic field. The magnetic field is produced using a coil with passage of time dependent current. A magnetostrictive material responds with a force, hence magnetostrictive actuator. The force produced should be used to counteract the forces due to vibration. Thus, damping is introduced. The idea of incorporting Terfenol-D layer facilitates the best control of the fluid conveying FGMT pipe to avoid the bifurcation and flutter instabilities and achieve more adaptive and efficient system. Additionally increasing or decreasing effect of feedback control gain and thermal loading on critical flow velocity and instabilities have been addressed.

Conflict of interest

The authors declare no conflict of interest.

Abbreviations

FGMT	Functionally graded material integrated with Terfenol-D
BC	boundary conditions
DQM	differential quadrature method
DTM	differential transform method
ODE	ordinary differential equation

Author details

Mukund A. Patil[1†] and Ravikiran Kadoli[2*†]

1 Department of Mechanical Engineering, G.H. Raisoni Institute of Engineering and Business Management, Jalgaon, Maharashtra, India

2 Department of Mechanical Engineering, National Institute of Technology Karnataka, Surathkal, Mangalore, Karnataka, India

*Address all correspondence to: rkkadoli@nitk.edu.in

† These authors contributed equally.

References

[1] Tijsseling AS. Fluid-structure interaction in liquid-filled pipe systems: A review. Journal of Fluids and Structures. 1996;**10**(2):109-146

[2] Ibrahim RA. Overview of mechanics of pipes conveying fluids—Part I: Fundamental studies. Journal of Pressure Vessel Technology. 2010;**132**(3):034001-1-32

[3] Ibrahim RA. Mechanics of pipes conveying fluids—part ii: Applications and fluidelastic problems. Journal of Pressure Vessel Technology. 2011;**133**(2):024001-1-30

[4] Li S, Karney BW, Liu G. Fsi research in pipeline systems–a review of the literature. Journal of Fluids and Structures. 2015;**57**:277-297

[5] Chen S-S. Free vibration of a coupled fluid/structural system. Journal of Sound and Vibration. 1972;**21**(4):387-398

[6] Zhai H-b, Zi-yan W, Liu Y-s, Yue Z-f. Dynamic response of pipeline conveying fluid to random excitation. Nuclear Engineering and Design. 2011;**241**(8):2744-2749

[7] Cigeroglu E, Samandari H. Nonlinear free vibrations of curved double walled carbon nanotubes using differential quadrature method. Physica E: Low-dimensional Systems and Nanostructures. 2014;**64**:95-105

[8] Tang Y, Yang T. Post-buckling behavior and nonlinear vibration analysis of a fluid-conveying pipe composed of functionally graded material. Composite Structures. 2018;**185**:393-400

[9] Li B, Wang Z, Jing L. Dynamic response of pipe conveying fluid with lateral moving supports. Shock and Vibration. 2018;**2018**:1-17

[10] Wang Y, Zhang Q, Wang W, Yang T. In-plane dynamics of a fluid-conveying corrugated pipe supported at both ends. Applied Mathematics and Mechanics. 2019;**40**(8):1119-1134

[11] Ni Q, Zhang ZL, Wang L. Application of the differential transformation method to vibration analysis of pipes conveying fluid. Applied Mathematics and Computation. 2011;**217**(16):7028-7038

[12] Li Y-d, Yang Y-r. Forced vibration of pipe conveying fluid by the green function method. Archive of Applied Mechanics. 2014;**84**(12):1811-1823

[13] Dai HL, Wang L. Dynamics and stability of magnetically actuated pipes conveying fluid. International Journal of Structural Stability and Dynamics. 2016;**16**(06):1550026

[14] Oke WA, Khulief YA. Effect of internal surface damage on vibration behavior of a composite pipe conveying fluid. Composite Structures. 2018;**194**:104-118

[15] Sutar S, Madabhushi R, Chellapilla KR, Poosa RB. Determination of natural frequencies of fluid conveying pipes using muller's method. Journal of The Institution of Engineers (India): Series C. 2019;**100**(3):449-454

[16] Rafiee R, Sharifi P. Stochastic failure analysis of composite pipes subjected to random excitation. Construction and Building Materials. 2019;**224**:950-961

[17] Rajidi SR, Gupta A, Panda S. Vibration characteristics of viscoelastic sandwich tube conveying fluid. Materials Today: Proceedings. 2020;**28**:2440-2446

[18] Amiri A, Masoumi A, Talebitooti R. Flutter and bifurcation instability analysis of fluid-conveying micro-pipes sandwiched by magnetostrictive smart layers under thermal and magnetic field. International Journal of Mechanics and Materials in Design. 2020;**16**(3):569-588

[19] Soltani P, Bahar P, Farshidianfar A. An efficient gdq model for vibration analysis of a multiwall carbon nanotube on pasternak foundation with general boundary conditions. Proceedings of the Institution of Mechanical Engineers, Part C: Journal of Mechanical Engineering Science. 2011;**225**(7):1730-1741

[20] Bahaadini R, Hosseini M. Effects of nonlocal elasticity and slip condition on vibration and stability analysis of viscoelastic cantilever carbon nanotubes conveying fluid. Computational Materials Science. 2016;**114**:151-159

[21] Askari H, Esmailzadeh E. Forced vibration of fluid conveying carbon nanotubes considering thermal effect and nonlinear foundations. Composites Part B: Engineering. 2017;**113**:31-43

[22] Kamali M, Mohamadhashemi V, Jalali A. Parametric excitation analysis of a piezoelectric-nanotube conveying fluid under multi-physics field. Microsystem Technologies. 2018;**24**(7):2871-2885

[23] Liang F, Yang XD, Bao RD, Zhang W. Frequency analysis of functionally graded curved pipes conveying fluid. Advances in Materials Science and Engineering. 2016;**2016**:1-9

[24] Zhou X-w, Dai H-L, Wang L. Dynamics of axially functionally graded cantilevered pipes conveying fluid. Composite Structures. 2018;**190**:112-118

[25] Dai J, Liu Y, Liu H, Miao C, Tong G. A parametric study on thermo-mechanical vibration of axially functionally graded material pipe conveying fluid. International Journal of Mechanics and Materials in Design. 2019;**15**(4):715-726

[26] Heshmati M. Influence of an eccentricity imperfection on the stability and vibration behavior of fluid-conveying functionally graded pipes. Ocean Engineering. 2020;**203**:107192

[27] Liang X, Zha X, Jiang X, Wang L, Leng J, Cao Z. Semi-analytical solution for dynamic behavior of a fluid-conveying pipe with different boundary conditions. Ocean Engineering. 2018; **163**:183-190

[28] Yi-min H, Ge Seng W, Wei, and He Jie. A direct method of natural frequency analysis on pipeline conveying fluid with both ends supported. Nuclear Engineering and Design. 2012;**253**:12-22

[29] Łuczko J, Czerwiński A. Three-dimensional dynamics of curved pipes conveying fluid. Journal of Fluids and Structures. 2019;**91**:102704

[30] Engdahl G, Mayergoyz ID. Handbook of Giant Magnetostrictive Materials. Vol. 386. San Diego: Academic Press; 2000

[31] Patil MA, Kadoli R. Differential quadrature solution for vibration control of functionally graded beams with terfenol-d layer. Applied Mathematical Modelling. 2020;**84**:137-157

[32] Reddy JN, Barbosa JI. On vibration suppression of magnetostrictive beams. Smart Materials and Structures. 2000; **9**(1):49-58

[33] Qian Q, Wang L, Ni Q. Instability of simply supported pipes conveying fluid under thermal loads. Mechanics Research Communications. Apr 2009; **36**(3):413-417

[34] Lin W, Qiao N. In-plane vibration analyses of curved pipes conveying fluid using the generalized differential quadrature rule. Computers & Structures. 2008;**86**(1–2):133-139

[35] Talebitooti M, Fadaee M. A magnetostrictive active vibration control approach for rotating functionally graded carbon nanotube-reinforced sandwich composite beam. Smart Materials and Structures. 2019;**28**(7): 075007

[36] Patil MA, Kadoli R. Effect of two-parameter partial foundation and viscoelastic supports on free vibration of Terfenol-D layered functionally graded fluid conveying pipe using domain decomposition technique. Mechanics of Advanced Materials and Structures. 2022:1-13. DOI: 10.1080/15376494.2022.2103604

[37] Li Li, Yujin Hu. Critical flow velocity of fluid-conveying magneto-electro-elastic pipe resting on an elastic foundation. International Journal of Mechanical Sciences. Dec 2016;**119**: 273-282

[38] Shamim M, Mehran S, Davood Y, Ebrahim E. Nonlinear vibration analysis of fluid-conveying microtubes. Nonlinear Dynamic. Mar 2016; **85**(2): 1007-1021

Chapter 7

Forced Laminar Flow in Pipes Subjected to Asymmetric External Conditions: The HEATT© Platform for Online Simulations

Mariano Alarcón, Manuel Seco-Nicolás, Juan Pedro Luna-Abad and Alfonso P. Ramallo-González

Abstract

This chapter studies the fluid flow within pipes subjected to thermal asymmetrical boundary conditions. The phenomenon at hand takes place in many real-world industrial situations, such as solar thermal devices, aerial pipelines. A steady-state analysis of laminar forced-convection heat transfer for an incompressible Newtonian fluid is studied. The fluid is considered to flow through a straight round pipe provided with straight fins. For the case studied, axial heat conduction in the fluid has been considered and the effects of the forced convection have been considered to be dominant. A known uniform temperature field is applied at the upper external surface of the assembly. The 3D assembly has been created combining cylindrical and Cartesian coordinates. The governing differential equation system is solved numerically through suitable discretization in a set of different finite volume elements. The results are shown through the thermal profiles in respect of longitudinal and radial-azimuthal coordinates and the problem characteristic length. To facilitate the resolution of this phenomenon, an open computing platform called HEATT©, based on this model, has been developed, and it is also shown here. The platform is now being built and is expected to be freely available at the end of year 2022.

Keywords: straight round pipe, asymmetrical boundary conditions, 3D simulation, conjugate-extended Graetz problem, HEATT platform

1. Introduction

Laminar forced convection mechanisms take place in many industrial installations where the processes involve the use of fluids in pipes. Numerous applications exist of such flows, including flat solar thermal collectors [1–4], solar trough devices [5], nanofluids [4, 6], mini and micro channels [7] and a wide variety of heat exchangers [8].

Currently, the interest on the studies of thermal behaviour in oil or gas pipelines is growing due to the international context, as well as the study of hydrogen through liquefied petroleum gases (LPG) pipelines that are already installed [9–12].

From the point of view of the analysis of fluids' heat exchange on pipelines, it is technically relevant to consider that flow and heat processes occur simultaneously, i.e. they are coupled, which increases the complexity of the process. Understanding flow behaviour under these conditions is key to pipeline design and device efficiency.

The fundamentals of the thermal mechanisms involved on fluid flow have been extensively studied [13–15]. The complexity in their analysis and the geometry of practical applications has made common its resolution using experimental and numerical studies, examples of them are [16, 17].

One of the more significant models of fluid behaviour within a pipe is the one known as Graetz Problem (GP), stated by Graetz in 1882 [18], where in a given point of the pipe a fluid flowing in laminar forced flow is subject to a sudden change in its external boundary conditions, either temperature or heat flow. Graetz proposed a bi-dimensional approach, neither considering pipe nor axial fluid conduction, which was analytically solved. More than a 100 years after his work, Graetz's problem continues to receive the attention of researchers. In the present century, some researchers have extended this known problem to take into account both the physical presence of the pipe and the axial fluid conduction (*conjugate extended Graetz problem*), finding their results by analytical [19] or numerical procedures [20, 21], also some have included transient process [22]. Other researchers have studied the flow with periodically varying inlet temperature in pipes of different shapes [23], pipes subjected to a sudden [24] or periodical change [25] in external heat or ambient temperature, etc. Also significant is the formulation of the concept of characteristic length of the process, carried out by discriminated dimensional analysis by Seco-Nicolás et al. [26]. All these studies, based on radial symmetry, assume the 2D hypothesis. However, in certain cases such relatively simplified models do not provide results as accurate as those obtained through tri-dimensional numerical models.

This chapter faces the problem of fluid flow within pipes subjected to thermal asymmetrical boundary conditions which take place in many real industrial situations such as those related to solar thermal devices, aerial pipelines subjected to external temperatures, etc. Other examples and attempts to solve this problem can be seen on: [27–31]. Despite the asymmetry of the problem, much simpler bi-dimensional models are currently used for pipe design purposes, ignoring the important consequences of the asymmetry that it exists.

This work presents a steady-state analysis of the laminar forced-convection heat transfer process for a liquid flowing through a straight round pipe when radially asymmetrical external conditions are applied to the tube's external surface (a known uniform temperature to the upper surface and adiabatic condition to the lower) and taking into consideration axial heat conduction in the fluid.

A governing differential equation system is coupled to the Laplace equation for the solid and is solved numerically through suitable discretisation in a set of different finite volume elements, considering the axial heat conduction in the fluid, but neglecting the heat generation by viscous dissipation, the buoyancy effects or the variation of the thermal properties of the materials.

Many techniques have been developed to simulate convective flows using finite element techniques [32], finite difference solvers [33], method of lines [34] and many others. In the present case to evaluate the proposed model Network Simulation Method (NSM) [35–37], a powerful numerical methodology has been chosen. This method, based on a finite-difference scheme, can virtually solve any ordinary and partial differential equation. The method starts from suitable discretisation of the problem [37–39]; after which an equivalent electric circuit of the process is built, including boundary conditions, the so-called network model. Finally, appropriate software is used to solve the circuit, from whose results the thermal response is obtained. NSM has been chosen in this case because it yields results that have been seen to be as accurate as those obtained with CFD software using significantly fewer computational resources. Open-source software NGSpice [40], originally created for electric circuit analysis, has been used in our case to solving this problem; also proprietary software such as Pspice© [41] can be used for this purpose. The model has been validated by experimental results obtained in an experimental solar thermal plant [42].

Related to this model, an open computing platform called HEATT©, based on this model and solved using the NTM, is now being built, which will allow online calculation of flow within pipes subject to complex thermal conditions. The platform is expected to be freely available to the public before the end of 2022.

2. Physical and mathematical model

The set-up studied here is composed by a flat plate welded to a round duct as shown in **Figure 1**. The flat plate acts as a fin. Its addition to the model is due to the fact that many of the pipes subjected to thermal stress incorporate fins to improve heat exchange.

Consider an isotropic fluid at uniform temperature, T_1, and a certain velocity profile inside a tube whose whole external surface is also at temperature T_1. At $z = 0$ the fluid enters a long duct, whose upper half surface is maintained from this point at a constant temperature T_2 ($T_2 > T_1$), while its lower surface is thermally isolated (**Figure 1**); these conditions are maintained along the whole duct length ($z > 0$). The plate is maintained at a constant temperature T_2 on its upper surface, and it is insulated on its lower surface and extremes in the same conditions of the pipe. Heat is transferred throughout the plate to the duct by conduction. Therefore, the plate acts as a fin.

Consequently, at $z = 0$, the fluid suffers a sudden change in its boundary conditions and heat is transferred through the tube by conduction, and then by convection to the fluid, in which conduction takes place in both radial and axial directions.

The fluid working conditions are similar to those of the bi-dimensional problem stated by Graetz [18], except for the radial asymmetry of the temperature boundary condition caused by the radially asymmetric thermal condition in the external surface

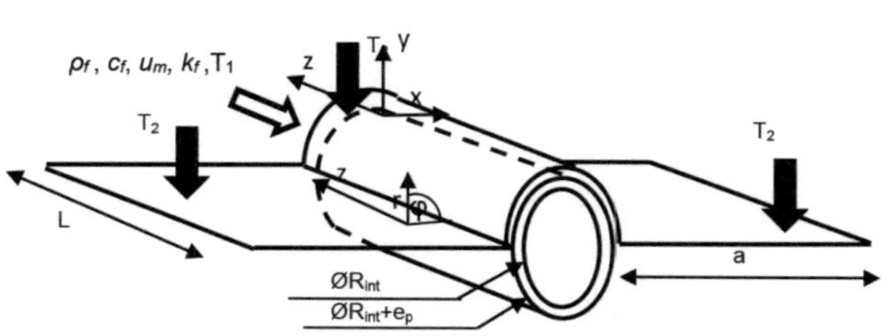

Figure 1.
Simulation model outline.

of the pipe and the presence of the fin. These boundary conditions make it necessary to formulate a 3D model.

The geometry of the problem requires incorporating both cylindrical and Cartesian coordinate systems. Regarding cylindrical coordinates for the round duct, direction z is located parallel to the axis of the pipe, the r vector is normal to it, and the third dimension is described by angle φ. Regarding Cartesian coordinate system for the fin, direction z is also located parallel to the axis of the pipe, and directions x and y are orthogonal to it, as can be seen in **Figure 2**.

As regards the material parts, the assembly consists of a round pipe, with k_s being the constant conductivity and e_p the constant thickness. The pipe is assumed straight and non-deformable. Heat generation by viscous dissipation or other sources is not considered. As regards the time domain, the model is considered to be at stationary state.

It is assumed that the fluid flows in laminar-forced convection in stationary regime, while its thermal properties (density, ρ_f, specific heat, c_f, and thermal conductivity, k_f) remain unchanged in their values at temperature T_1.

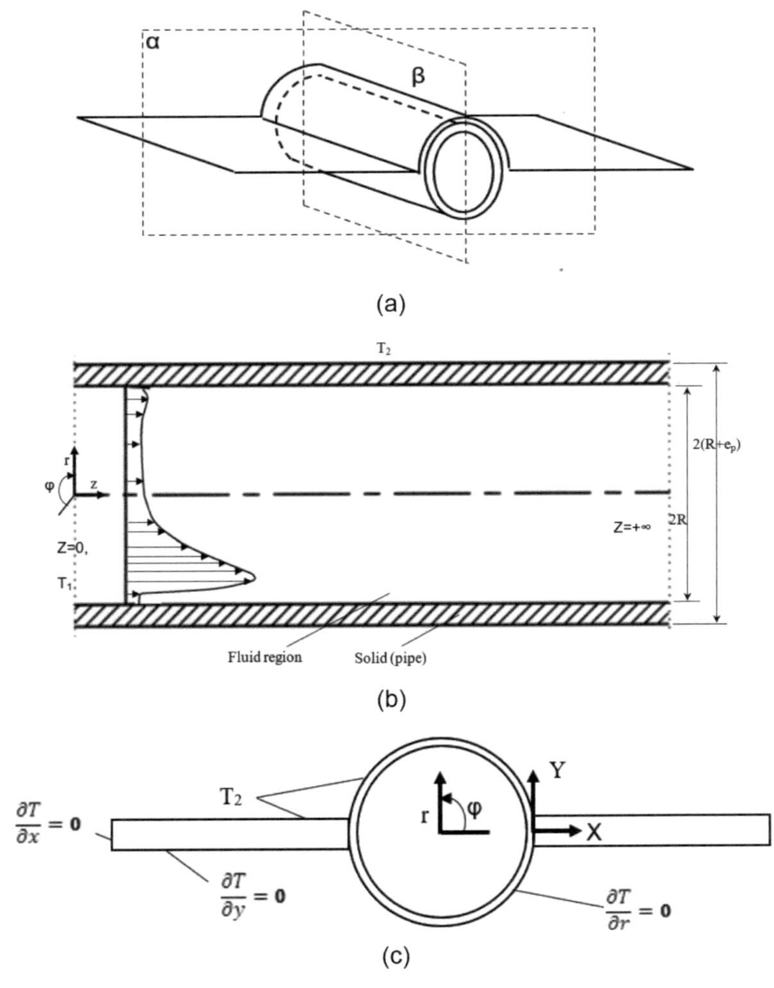

Figure 2.
3D Duct problem geometry. (a) Section planes outline. (b) 2D longitudinal section β including velocity profile. (c) 2D transversal section α.

For the velocity profile, a polynomic profile of grade 10 has been chosen. This kind of profiles can be found in certain devices, such as inclined tubes [43]. The velocity will be considered invariable along the whole pipe (**Figure 2b**).

The viscous buoyancy-driven heat transfer is considered negligible due to dominant effects of the studied forced convection in a laminar and incompressible Newtonian fluid flow, which is the case of water and other fluids in certain conditions [44].

Under these conditions, the tri-dimensional equations that govern the coupled system are (on cylindrical coordinates) [45, 46]:

Equation of the solid (pipe) region. Cylindrical coordinates:

$$\frac{1}{r}\left[\frac{\partial}{\partial r}\left(rk_s\frac{\partial T_s}{\partial r}\right)\right]+\frac{1}{r^2}\frac{\partial}{\partial \varphi}\left(k_s\frac{\partial T_s}{\partial \varphi}\right)+\frac{\partial}{\partial z}\left(k_s\frac{\partial T_s}{\partial z}\right)=(\rho c)_s\frac{\partial T_s}{\partial t} \tag{1}$$

Equation of the fluid (inside pipe) region. Cylindrical coordinates:

$$\frac{1}{r}\left[\frac{\partial}{\partial r}\left(rk_f\frac{\partial T_f}{\partial r}\right)\right]+\frac{1}{r^2}\frac{\partial}{\partial \varphi}\left(k_f\frac{\partial T_f}{\partial \varphi}\right)+\frac{\partial}{\partial z}\left(k_f\frac{\partial T_f}{\partial z}\right)=(\rho u_z c)_f\frac{\partial T_f}{\partial z}+(\rho c)_f\frac{\partial T_f}{\partial t} \tag{2}$$

Equation of the fin. Cartesian coordinates:

$$\frac{\partial}{\partial x}\left(k_a\frac{\partial T_a}{\partial x}\right)+\frac{\partial}{\partial y}\left(k_a\frac{\partial T_a}{\partial y}\right)+\frac{\partial}{\partial z}\left(k_a\frac{\partial T_a}{\partial z}\right)=(\rho c)_a\frac{\partial T_a}{\partial t} \tag{3}$$

This model can be considered as an extension of the classical conjugate-extended Graetz problem and has been widely used in the literature [20–22].

Time-dependent terms will be omitted in the current model due to fact that the study focuses on the stationary phenomenon, and conduction coefficients are considered equal in all directions as the fluid is considered isotropic and with invariable thermal properties. The rest of boundary conditions that define the problem are detailed in **Table 1**.

At $z = 0$, $r < R + e_p$	$T_f = T_s = T_1$; $\partial T_a/\partial z = 0$	(4)
At $z = +\infty$	$\partial T_f/\partial r = \partial T_s/\partial r = 0$ $\partial T_a/\partial z = 0$	(5)
At $r = R$	$T_f = T_s$ $(k_s/k_f)\partial T_s / \partial r = \partial T_f / \partial r$	(6)
At $r = R + e_p$, $\varphi \in (0, \pi]$	$T_s = T_a = T_2$	(7)
At $r = R + e_p$, $\varphi \in 0$	$T_s = T_a$	(7 bis)
At $r = R + e_p$, $\varphi \in (\pi, 2\pi)$	$\partial T_s/\partial r = 0$	(8)
At $y = e/2$	$T_a = T_2$	(9)
At $y = -e/2$	$\partial T_a/\partial y = 0$	(10)
At $x = a$	$\partial T_a/\partial x = 0$	(11)

Table 1.
Analytical and dimensional boundary conditions.

3. Numerical model

Numerical solutions of coupled governing Eqs. (1)–(3) under boundary conditions (4)–(11) have been reached using Network Simulation Method (NSM) and the circuit solver NGSpice.

3.1 The network model

In NSM terminology, the equivalent, or analogous, electrical circuit of the considered process is called the *network model*. The construction of the network model begins with the discretisation of the mentioned system of differential equations in a finite volume elements mesh which transforms the governing partial differential equations of balance Eqs. (1)-(3) into the set of algebraic Eqs. (12)–(14).

Discretisation of the equation of the solid (duct) region. Cylindrical coordinates:

$$\frac{1}{r}\cdot\left[\frac{T_{j+\Delta r/2}-T_{j-\Delta r/2}}{\Delta r}\right]+\frac{1}{\Delta r}\cdot\left[\frac{T_{j+\Delta r/2}-T_{j-\Delta r/2}}{\Delta r}\right]+\frac{1}{r^2\varphi}\left[\left(T_{\varphi+\Delta\varphi/2}-T_{\varphi-\Delta\varphi/2}\right)/\Delta\varphi\right]$$
$$+1/\Delta z\cdot\left[\left(T_{i+\Delta z/2}-T_{i-\Delta z/2}\right)/\Delta z\right]$$
$$=0 \tag{12}$$

Discretisation of the equation of the fluid (duct) region. Cylindrical coordinates:

$$\begin{aligned}&1/r\cdot\left[\left(T_{j+\Delta r/2}-T_{j-\Delta r/2}\right)/\Delta r\right]+1/\Delta r\cdot\left[\left(T_{j+\Delta r/2}-T_{j-\Delta r/2}\right)/\Delta r\right]\\&+1/r^2 1/\varphi\cdot\left[\left(T_{\varphi+\Delta\varphi/2}-T_{\varphi-\Delta\varphi/2}\right)/\Delta\varphi\right]+1/\Delta z\cdot\left[\left(T_{i+\Delta z/2}-T_{i-\Delta z/2}\right)/\Delta z\right]\\&-((\rho\cdot c\cdot u_z)/k)_f\left[\left(T_{i+\Delta z/2}-T_{i-\Delta z/2}\right)/\Delta z\right]\\&=0\end{aligned} \tag{13}$$

Discretisation of the equation of the fin. Cartesian coordinates:

$$\begin{aligned}&1/\Delta x\cdot\left[\left(T_{i+\Delta x/2}-T_{i-\Delta x/2}\right)/\Delta x\right]+1/\Delta y\cdot\left[\left(T_{i+\Delta y/2}-T_{i-\Delta y/2}\right)/\Delta y\right]\\&+1/\Delta z\cdot\left[\left(T_{i+\Delta z/2}-T_{i-\Delta z/2}\right)/\Delta z\right]\\&=0\end{aligned} \tag{14}$$

Based on the governing equations and boundary conditions, an electrical circuit for each equation has been created (**Figure 3**). Each term in these equations becomes an

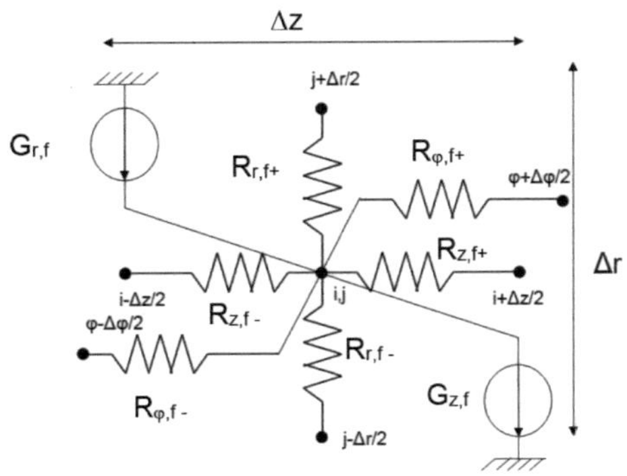

Figure 3.
Electrical circuit scheme of the fluid basic cell.

electric component. Consequently, each finite volume element (elementary cell) is composed of a set of electrical elements according to the thermo-electric analogy corresponding to the different terms of the previous equations. In this temperature is equivalent to voltage, and heat fluxes ($\partial T/\partial x$, $\partial T/\partial y$, $\partial T/\partial z$, $\partial T/\partial r$ and $\partial T/\partial \varphi$) are equivalent to electric currents. More details of the fundamentals of the NSM can be found in Ref. [42].

From the point of view of the NSM, the terms contained in each of the above equations can be considered as currents according to the currents Kirchhoff's law (because its summation over a node needs to be zero). As an example, the term $1/r \cdot \left[\left(T_{j+\Delta r/2} - T_{j-\Delta r/2}\right)/\Delta r\right]$ introduces heat conduction in cylindrical co-ordinates both on the fluid and on the pipe equation. The step-by-step description of the elementary cell building (**Figure 3**) has been detailed in the Appendix of this chapter.

Eq. (13) describes the three-dimensional flow behaviour which is dominated by the axial velocity u_z, assumed as a velocity function of order 10, profile equation of which is done by Eq. (15):

$$\begin{aligned} u_z = {} & 1.06 \cdot 10^{-2} + \left(\frac{r}{R}\right) \cdot 4.59 \cdot 10^{-1} + 3.75 \cdot \left(\frac{r}{R}\right)^2 + 19.4 \cdot \left(\frac{r}{R}\right)^3 + 15.6 \cdot \left(\frac{r}{R}\right)^4 \\ & - 92.1 \cdot \left(\frac{r}{R}\right)^5 + 40.6 \cdot \left(\frac{r}{R}\right)^6 + 843 \cdot \left(\frac{r}{R}\right)^7 + 577 \cdot \left(\frac{r}{R}\right)^8 - \left(\frac{r}{R}\right)^9 \cdot 3.26 \cdot 10^3 \\ & - \left(\frac{r}{R}\right)^{10} \cdot 4.93 \cdot 10^3 \end{aligned} \tag{15}$$

In total, the entire system has been discretised using a three-dimensional mesh of identical cells in a manner that the square-sectioned straight fin is divided into 200 cells in z direction, 10 cells in the x direction and 1 cell in the y direction (**Figure 4**). The pipe is also symmetrically divided into 200 cells in the z direction, 4 cells in φ direction and 7 cells in the r direction where 5 belongs to the fluid and 2 to the pipe thickness (**Figure 4**). This mesh has been considered to be sufficient for the purpose of this work.

Figure 5 shows the five main planes of the pipeline (named similarly to the cardinal points) that will be referred to in Section 4 to identify the different points of the pipeline's cross section. Although we have pointed out the asymmetry of the physical model, it should be noted that the pipe-fin assembly is radially asymmetric, but it presents symmetry with respect to the N-S plane.

Once the elementary circuit has been built, the boundary conditions must be implemented. In this case, some of the most relevant are voltage sources than fit the constant external temperature T_2 in the upper side of the assembly and infinite (very

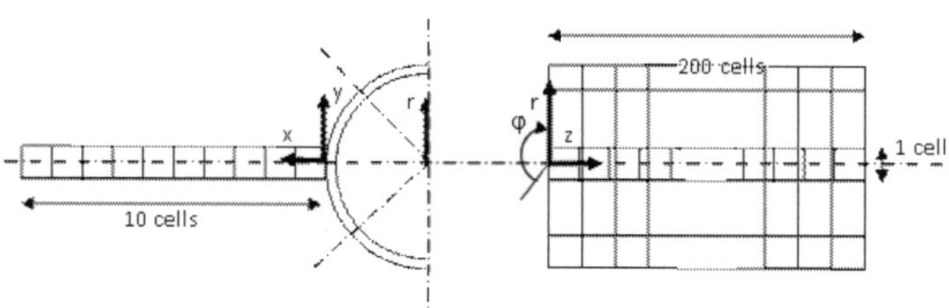

Figure 4.
Numerical model discretisation mesh.

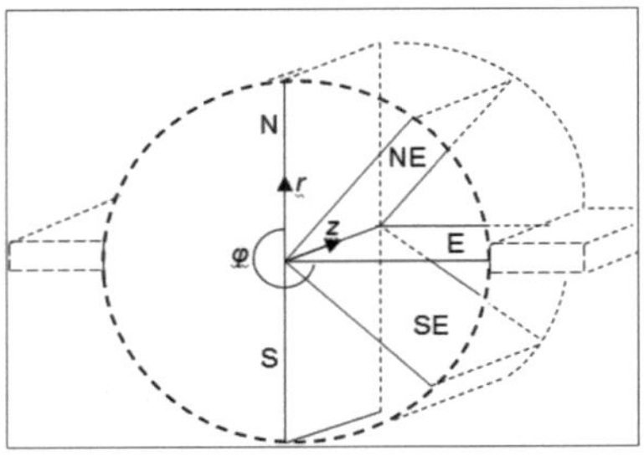

Figure 5.
Longitudinal pipe section planes screened in simulations.

high) resistance in its lower side. Finally, elementary cells and devices corresponding to the boundary conditions are assembled building an electric circuit (network model), which can be solved using an appropriate software, such as NGSpice.

A complete description of the circuit construction as well as computing details of the simulation can be found in the Appendix and reference [42]. In order to know the numerical value of the different electrical components forming out network (resistors, voltage and current sources, etc.), it is necessary to build the circuit and to obtain realistic values of parameters and boundary conditions corresponding to the case study. In our case, a solar thermal collector. The parameters are listed on **Table 2**. As stated above, solar thermal devices are one of the practical applications of this model, with obvious similarities in their geometry.

In this case, when Reynolds number is high (41989) and Rayleigh number remains low ($3.92 \cdot 10^3$), the fluid flows can be considered in a laminar forced-convection regime, and in consequence, the buoyancy effects are negligible [44]. The minimum pipe length in which the thermal phenomena are developed came from the use of the Nusselt number approximations [47].

About the length of the pipe, it has been used the concept of characteristic length, l*, defined as the length needed for the fluid to fully develop the thermal process [26], in

Variable	Value	Units
ρ_f	982.3	$kg \cdot m^{-3}$
c_f	$4.207 \cdot 10^3$	$J \cdot kg^{-1} \cdot K^{-1}$
R_{in}	$3.5 \cdot 10^{-3}$	m
e_p	$5 \cdot 10^{-4}$	m
k_f	0.66	$W \cdot m^{-1} \cdot K^{-1}$
k_s	380	$W \cdot m^{-1} \cdot K^{-1}$
μ	$4.27 \cdot 10^{-4}$	$kg \cdot m^{-1} \cdot s^{-1}$
L	4	m
T_1	65.45	°C
T_2	100	°C
a	$1.2 \cdot 10^{-2}$	m
e	$5 \cdot 10^{-4}$	m

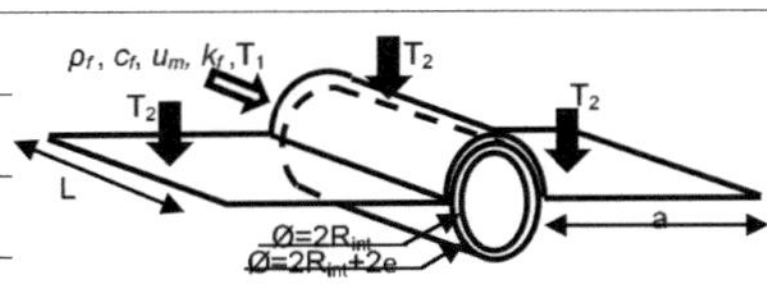

Table 2.
Simulated model magnitudes.

this case to achieve the external temperature T_2. Therefore, the simulated pipe should be long enough ($L > l^*$), to ensure that the thermal phenomena due to the sudden change in temperature at $z = 0$ has ended. This *hidden parameter* must be previously determined. In this case, the preliminary simulation with a 2D model guided the determination of the characteristic length (afterwards checked). So, we found that the corresponding characteristic length in this case would be 2 m if the whole exterior pipe surface was kept isothermally surrounded. This is not exactly the case, because in the 3D model presented in this chapter, only upper half surface is at constant temperature and the rest is insulated; consequently, 4 m length duct has been simulated.

3.2 Model validation

The 3D numerical model built following these rules and using the parameters mentioned in the previous section was validated by comparing its results to literature and experimental data [42].

The case modelled in the present chapter is a complete innovation; therefore, no references for comparison can be found in the literature. Furthermore, no references of the 3D configuration of the conjugate extended Graetz problem could be found in the literature previously to the work of reference [42] of the same authors of this chapter. In comparison to the finned tube model of the present chapter, the above cited reference modelled a bare tube (without fins) with a parabolic velocity profile, both being models analogous in the rest of characteristics.

Figure 6 shows the comparison of the fluid temperature field at 0.5 m of the entrance in both models. Main differences are a consequence of the different fluid velocity and the presence of fins. **Figure 6a** depicts the isothermal lines of the finned tube, where a slight tendency towards horizontality can be appreciated in the surroundings of the tube-fin junction compared with that of the **Figure 6b** of the bare tube. Nevertheless, the bulk fluid is little affected by the presence of fins. This was to be expected as in the finned tube, the temperature of the fins is the same as that of the upper half-pipe in both models. Consequently, it can be said that the fluid temperature field of the finned tube model agrees with that of the bare tube, and it can be considered that the validation procedure carried out in Ref. [42] applies to the present model.

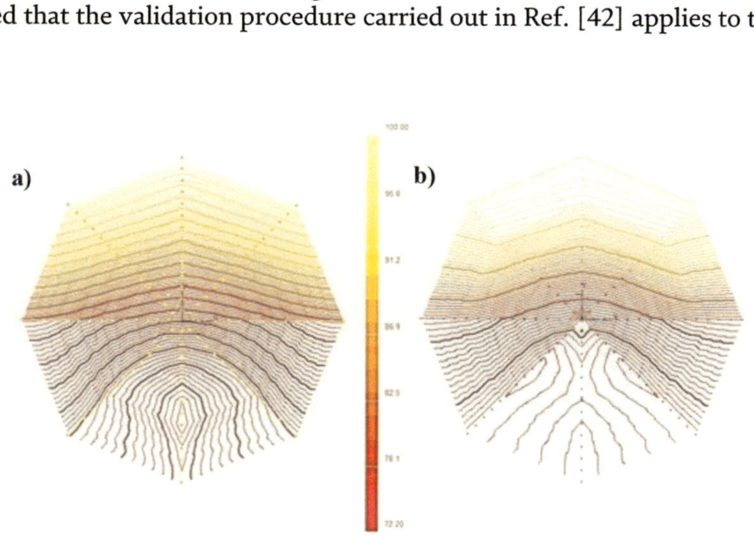

Figure 6.
Comparison of the cross-section temperature maps at 0.5 m in the finned tube (a) and in a bare tube (b).

In respect to the code, the results of the 3D model "acting as" a 2D model (radially symmetric boundary conditions) were compared with the 2D published results [21, 22], obtaining relative errors along the whole tube length below 2% and mean of 0.98%, the error standard deviation [42] being 0.56%. Analysis of the typified residuals of the 2D and 3D simulation results, error variance and a regression analysis were carried out. The coefficient of skewness (nearly 0) and kurtosis showed that the error data set was normally distributed. Finally, the relative error of the 3D model acting as a 2D model, and the 2D model was found to be 0.98 ± 7.74E-2% at a 95% confidence level.

On the other hand, the external temperatures (tube) of the cross section of the 3D simulation at different lengths were compared with the experimental measures yielded in a solar thermal experimental rig [42]. An analogous error study was conducted between numerical and experimental temperature data, concluding that the 3D numerical results were sufficiently close to those measured experimentally, relative errors being of 3.40 ± 0.601%, at 95% of confidence. More details of experimental rig, measured data and error study can be found in [42].

Consequently, those results confirmed the accuracy of the bare tube 3D model, and this conclusion can be extended to the finned tube model of this chapter, substantially equivalent to that of reference [42], especially as regards the fluid, and which can be considered as a continuation of the bare tube model.

4. Results and discussion

Typical fluid flow in laminar regime shows a parabolic profile. Nevertheless, there are situations in which velocity has a more complex profile [43]. Because of that a polynomial function of grade 10 (**Figure 7**) has been used in this work. Eq. (15) is the formula corresponding to this curve, and it is the value for u_z in Eqs. (2) and (13). Eq. (15) has been yielded from experimental results, and it corresponds to the case of the flow in tilted solar collectors [43].

4.1 Axial temperature profile and characteristic length of the process

Figure 8 shows the profile of the fluid temperature along the pipe in the different planes of the pipe (see **Figure 4**). In this case, the temperature rises faster than in the case of the parabolic profile of the velocity, reaching the external temperature T_2 even

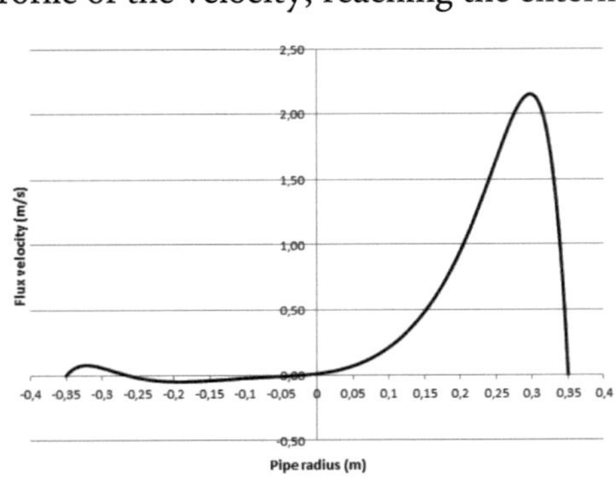

Figure 7.
Axial velocity profile following [43].

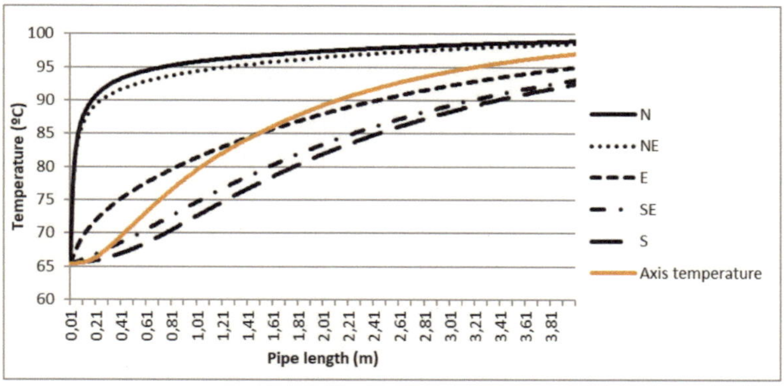

Figure 8.
Temperature profiles for u_z given by Eq. (15).

in the lower sections of the tube, i.e. flows with this speed profile mix much faster than with the laminar parabolic profile, yielding a higher convective coefficient.

If characteristic length, l*, is defined as the one where the temperature in the middle of the pipe (red line) reaches 90% of the total temperature jump, in this case 65.45 + 0.9·(100–65.45) = 96.55°C, from **Figure 7** gives l* = 3.6 m. Beyond this point, the temperature of the fluid does not increase substantially, which indicates that the rest of the pipe is not efficient for heat transfer purposes, so the length could be reduced.

4.2 Fluid temperature maps in a cross section

Figure 9 shows the temperature field of the cross section of the fluid (perpendicular to the axis) and of some vertical and horizontal sections of it at 13 cm from the inlet. As expected, the temperature curves show symmetry about the vertical axis.

Figure 9a represents the thermal map of the cross section. It can be observed that the fluid initially (z = 0) at T_1 (65.45°C) acquires at 13 cm from the entrance temperatures close to T_2 (100°C) in the layers near the top of the pipe (dark brown layers between the N ($\varphi = \pi/2$) and NE ($\varphi = \pi/4$) planes and nearby areas). At this point of the pipe, the upper half of the fluid shows almost parallel isotherms following the circular curve of the pipe, whose values decrease from plane N, (at the top) to the horizontal plane (plane E, $\varphi = 0$) and even up to almost plane SE ($\varphi = -\pi/4$). Sections A-A' and D-D′ (**Figure 8b** and **e**, respectively) clearly show this situation, presenting a minimum at the N-S axis as a consequence of the shape of the isotherms. Compared with the case of velocity parabolic profile [42], it can be said that in the case of polynomial profile, the temperature increases much faster than in the former, reaching the external temperature T_2 in the lower sections of the tube relatively close to the entrance, i.e. flows with this speed profile mix much faster than with the laminar parabolic profile, yielding a higher convective coefficient. A consequence of this is the shorter characteristic length shown by the polynomial velocity process compared with that of parabolic velocity profile.

In contrast, near the south (S) and southeast (SE) planes, the isothermal curves take a U-shape, except near the N-S axis, where a loop is formed at 68°C approximately in the centre of the bottom half of the tube. This is due to the fact that the fluid is being heated from the top half of the tube (whose temperature has been imposed at a uniform value of T_2), while the lower half of the tube, which is externally insulated,

a) Map of temperature of the cross section

b) Section A-A′

c) Section B-B′

d) Section C-C′

e) Section D-D′

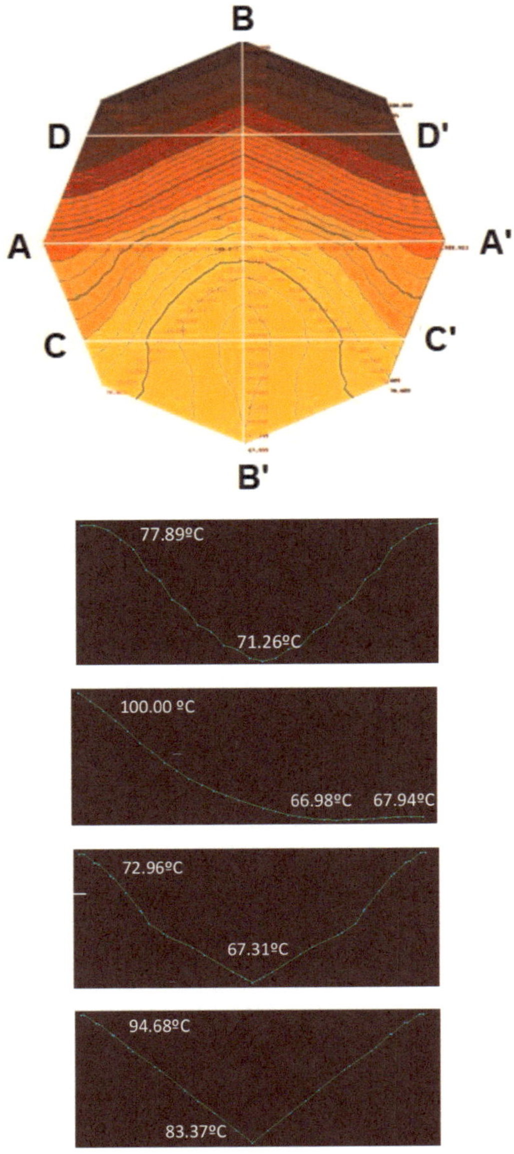

Figure 9.
Fluid temperature map of the cross section (a) and temperature profiles for different sections (b–e) at 13 cm from the entrance of the pipe.

is warmed by heat conduction from the solid pipe due to the fact that $k_f >> k_s$. As a result, in the S plane ($\varphi = -\pi/2$), the fluid near the tube is hotter than in the middle of the tube. This fact can also be detected in Section B-B′, which corresponds to the N-S plane (**Figure 9c**), where the minimum (66.98°C) is not found in the lower part of the section; you can see that the temperature profile is roughly flat along the bottom quarter of the section. This suggests that the thermal process is more dependent on pipe-wall conduction effects than on the velocity profile, even with such a complex and irregular profile, due to the low conductivity of the fluid and the laminar flux. It is relevant to note that the classic Graetz Problem does not consider the pipe through which the fluid flows, thus making it impossible to detect this temperature distortion.

Another issue is the influence of the fin on fluid temperatures. The fin-tube junction takes place in plane E, which corresponds to section A-A' (**Figure 9b**). A small distortion of the isotherms can be observed at this point compared with the case of a bare duct [42]. In this case, this distortion is not very important because the external temperature T_2 is uniform, both in the pipe and in the fin, and a very thin fin with little heat conduction capacity has been modelled. Otherwise, it could be very relevant, as in the case of the heat boundary condition.

Although not studied in this work, consequences from the non-uniform temperature field, which affects most of the thermo-fluid properties (density, viscosity, etc.) among others, can be drawn. This behaviour could impact on the operation of the pipeline and the related equipment.

5. The HEATT© platform

The methodology explained above has been used to develop a 3D model to simulate a laminar flow under conditions of forced convection subject to asymmetric boundary conditions, such as those found in the tube grid of a low-temperature solar thermal collector, in which the upper half of each tube receives energy from the sun, while the bottom half of the tube remains embedded in a layer of insulation [42]. The simulated results could be compared with the previous bibliography and with the experimental results obtained from the Solar Laboratory of the University of Murcia under real operating conditions, obtaining great differences between the results obtained using 2D models versus 3D models. The model roughly coincides with the one presented in this chapter (except for the fact of considering tube without fins).

This model required the creation of a three-dimensional mesh with nearly 17,000 cells, in which an electrical equivalent circuit of approximately 140,000 elements was implemented. The model includes resistors, voltage and current sources and capacitors.

Once the problem has been solved for a sufficiently variety of real cases, the tool has been created for solving the problem using a web as a service, providing free service to any professional or researcher anywhere in the world without the need to acquire expensive software or to instal any application that becomes obsolete with the evolution of the Operating Systems.

The resulting web application has been named HEATT ©, acronym for Pipeline Thermal Analysis and Assessment Tool (in Spanish: Herramienta de Evaluación y Análisis Térmico de Tuberías).

This simple and friendly platform is currently being launched in its version 1.0 as a Proof of Concept, to be released "as a service" for general use and for researchers and professionals from all over the world to send their opinions and improvement proposals to gradually make it growing up.

6. Conclusions

A numerical physical–mathematical model is presented for a laminar forced-convection fluid flow within a finned round duct subjected to constant temperature on its upper side and insulated on its lower side. The governing heat transfer and fluid flow 3D partial differential equations, combining cylindrical and Cartesian coordinates, are solved for steady-state conditions. The numerical model is run using the accurate contrasted Network Simulation Method, a low-workload computational method.

In the present chapter, the Graetz Problem is extended to incorporate axial fluid conduction, 3D coordinates, wall thickness with attached longitudinal flat-fins, radially asymmetrical boundary conditions and highly non-linear velocity profiles. Realistic conditions corresponding to the flow in flat plate solar collectors have been used in the simulations. A high non-linearity velocity profile has been evaluated, confirming that solid conductivity and thickness effects are not negligible for the studied thermal phenomena.

Temperature evolution across the fluid is analysed in detail. Different temperature values were found for different angles within every plane, due to the radial asymmetry of the geometry. Temperature fluid maps were obtained at different distances from the entrance of the pipe. Isotherms show parallel-like shapes on the top half of the tube, which become distorted in the lower half, where some loops appear due to the conduction effects of the studied pipe wall thickness, illustrating the non-uniformity of the temperatures within the fluid.

In addition, 3D simulation reveals that, in cases of asymmetry, the thermal phenomena require much more length to completely develop the flow than the length yielded by the 2D radially symmetric model. This is relevant because the 3D simulation reveals that the pipe needs up to six times more length than that predicted by 2D model.

When the characteristic length of the problem was considered (a virtual dimension equivalent to the distance at which the thermal phenomena are fully developed), the 2D approach was found to be no longer valid. That occurs in most of the studied real cases, when the pipe is subjected to asymmetrical boundary conditions.

All those changes enhance the solution of the Graetz problem and bring it nearer to real pipe conditions. From this, a basis for future works, including heat boundary conditions, different union thermal resistances or a variety of complex velocity functions, among many other possibilities, may be mentioned in order to encourage future optimisation studies. Meanwhile, the findings of the present study have applications in solar energy collectors, thermal heat dissipators, oillines and heat exchangers among many other facilities.

In addition, an open computing platform called HEATT©, based on this model, is now being built. The platform is expected to be freely available to the public before the end of 2022.

Acknowledgements

This project has been possible thanks to the funds obtained in public concurrence in the 2021 call for Proofs of Concept from the Seneca Foundation – Agency for Science and Technology of the Region of Murcia (Spain). Ph.D, Ramallo-González would like to thanks the European Commission for their funding of project PHOENIX grant number 893079.

Conflict of interest

The authors declare no conflict of interest.

Nomenclature

a absorber width (m)
c specific heat ($J \cdot kg^{-1} \cdot K^{-1}$)

e	absorber thickness (m)
e_p	pipe thickness (m)
G	voltage-dependent current source (A)
k	thermal conductivity (W·m^{-1}·K^{-1})
J	electric current (A)
L	pipe/absorber length (m)
l^*	characteristic length (m)
R	inner radius of the pipe (m)
R	Resistor (Ohm)
r	radial coordinate
T	Temperature (°C)
t	Time (s)
T_1	temperature in the entrance of the fluid (°C)
T_2	temperature in the exit of the fluid (°C)
u	Velocity (m·s^{-1})
z	axial coordinate (m)

Superscripts

$'$	dimensionless variable

Subscripts

a	associated to absorber (fin)
ext	exterior
f	associated to fluid
$i\,j, k, \varphi, l$	discretisation counters
in	inner
m	medium/average
p	solid–liquid interface
r	associated to radius
s	associated to solid
z	associated to z axis
φ	assoc. to azimuthal coordinate

Greek characters

φ	azimuthal coordinate (rad)
μ	dynamic viscosity (kg·m^{-1}·s^{-1})
ρ	density (kg·m^{-3})

A. Appendix. Procedure of building the numerical model

The procedure that needs to be followed to reproduce the simulation results of this chapter is described below.

A.1 Discretisation of the spatial domain

In this case, we can account three different regions or sub-domains, as stated in the physical–mathematical model, every one described by its corresponding equation of thermal behaviour: solid (pipe) region, Eq. (1), fluid (inside pipe) region, Eq. (2) and the fin, Eq. (3). Each region is divided into an adequate number of bi-dimensional cells (the third dimension is the axial coordinate, which is common for all cells). In this problem, the number of discrete dimensions are:

a. Pipe: Two divisions in the radial coordinate and four divisions in the azimuthal coordinate (**Figure 5**). Taking into account the symmetry of the problem in respect to the N-S plane, only half section must be simulated.

b. Fluid: Five divisions in the radial coordinate and, in accordance to the pipe, four divisions in the azimuthal coordinate.

c. Fin: Ten divisions of the x-dimension and one (no discretisation) in the y-dimension (**Figure 4**).

The third dimension, i.e. z-dimension, has been divided into 200 parts (**Figure 4**) and is common for all the bidimensional cells previously accounted for.

In summary, an overall number of 38 divisions (8 for the pipe, 20 for the fluid and 10 for the fin) of the cross section have been done, which gives a total of 38x200 = 7600 cells for the whole domain. In this case, the results have shown that the discretisation is adequate.

A.2 Building the numerical model

This step is divided into different minor steps, each one has to be applied to the corresponding region. Due to the fact that the most complex region is that of the fluid, and in order not to be repetitive, the process of construction of its elementary cell. The starting point is the set of discretised Eqs. (12)–(14) of the problem.

A.3 Building the elementary cell

The Network Simulation Method is based in the well-known thermal-electrical analogy [35]. In accordance to this, temperature is equivalent to voltage, and $\Delta T/\Delta z$, $\Delta T/\Delta r$ and $\Delta T/(r\Delta\varphi)$, which are related with heat fluxes, are equivalent to electric currents. Each of the Eqs. (12)-(14) represents the energy balance in the corresponding cell. These cells, at steady state must have a zero summation of currents. Consequently, each term of the differential equation must be converted into a current of the elementary circuit, which as a whole fulfils Kirchhoff's Laws.

In order to arrange the equations in a more convenient way, Eqs. (12) and (13) are multiplied by $\Delta z \cdot \Delta r \cdot \Delta\varphi$; Eq. (14) is multiplied by $\Delta x \cdot \Delta y \cdot \Delta z$. Note that the elementary cell (**Figure 3**) is in fact divided in two parts, being the node at the centre of the cell.

Tables 3, **4** and **5** contain the formulae of the different devices that integrate the elementary cells of solid pipe, fluid and fin, respectively. As the most complex, the fluid region elementary circuit is explained. In this case, five terms are found:

a. J_1 is a non-lineal term which comes from the increasing cell area with the radius. It must be implemented by a voltage-dependent current source, $G_{r,f}$ (**Figure 3**), value of which is expressed by Eq. (15), as it can be seen in **Table 4**.

$$J_1 = \frac{\Delta z \cdot \Delta\varphi}{r}\left[\left(T_{j+\frac{\Delta r}{2}} - T_{j-\frac{\Delta r}{2}}\right)\right] \tag{16}$$

b. The second term of the Eq. (13) is a current which corresponds to the heat flux in r-direction, $T_{j+\Delta r/2}$ and $T_{j-\Delta r/2}$ being the temperatures at both sides of the elementary j-cell in this direction, which are made equivalent to the external voltages of the elementary circuit, Eq. (16).

$$J_2 = \Delta z \cdot \Delta\varphi\left[\frac{\left(T_{j+\frac{\Delta r}{2}} - T_{j-\frac{\Delta r}{2}}\right)}{(\Delta r/2)}\right] \tag{17}$$

Consequently, the thermal resistance is represented by two electric resistances of value $R_{r,f} = \frac{\Delta r}{2\Delta z \cdot \Delta\varphi}$, as can be seen in **Table 4**. Similarly, the third and fourth terms become J_3 and J_4 currents corresponding to heat flux in the φ and z directions. The value of the thermal resistances $R_{\varphi,f}$ and $R_{z,f}$ yielded from these terms can be found in **Table 4**.

c. Finally, the fifth term represents the axial heat conduction due to the velocity field and has been implemented by a voltage-dependent current source, $G_{z,f}$, whose value is given by Eq. (17):

$$J_5 = \Delta z \cdot \Delta\varphi((\rho \cdot c \cdot u_z)/k)_f \cdot \left(T_{i+\Delta z/2} - T_{i-\Delta z/2}\right) \tag{18}$$

The location and connections of the different electric devices in the fluid elementary cell can be seen in **Figure 3**; plus or minus symbols in the different resistors distinguish both sides of the cell in respect to the central node.

This procedure must be followed for the different regions of the assembly, i.e. the fluid, the pipe and the fin. The pipe equation, Eq. (12) has only four terms, all of them coincide with those of the fluid except for the velocity term, and the fin, Eq. (14), has

Term	Discretisation of equations	Thermal-electrical analogy	Type of device
1	$\frac{\Delta\varphi \bullet \Delta z}{r}\left[T_{j+\frac{\Delta r}{2}} - T_{j-\frac{\Delta r}{2}}\right]$	$G_{r,s} = \Delta T \bullet \frac{\Delta z \cdot \Delta\varphi}{r}$	Voltage-dependent current source
2	$\Delta\varphi \bullet \Delta z \cdot \left[\frac{T_{j+\frac{\Delta r}{2}} - T_j}{\Delta r_s/2} - \frac{T_j - T_{j-\frac{\Delta r}{2}}}{\Delta r_s/2}\right]$	$R_{r,s} = \frac{\Delta r_s}{2\Delta z \cdot \Delta\varphi}$	Resistor
3	$\frac{\Delta r \bullet \Delta z}{r^2} \cdot \left[\frac{T_{\varphi+\frac{\Delta\varphi}{2}} - T_\varphi}{\Delta\varphi/2} - \frac{T_\varphi - T_{\varphi-\frac{\Delta\varphi}{2}}}{\Delta\varphi/2}\right]$	$R_{\varphi,s} = \frac{r^2 \cdot \Delta\varphi}{2\Delta r_s \cdot \Delta z}$	Resistor
4	$\Delta r \bullet \Delta\varphi\left[\frac{T_{i+\frac{\Delta z}{2}} - T_i}{\Delta z/2} - \frac{T_i - T_{i-\frac{\Delta z}{2}}}{\Delta z/2}\right]$	$R_{z,s} = \frac{\Delta z}{2\Delta r_s \cdot \Delta\varphi}$	Resistor

Table 3.
Formulae and analogous electrical devices yielded from solid Eq. (12).

Term	Discretisation of equations	Thermal-electrical analogy	Type of device
1	$\frac{\Delta\varphi \bullet \Delta z}{r}\left[T_{j+\frac{\Delta r}{2}} - T_{j-\frac{\Delta r}{2}}\right]$	$G_{r,f} = \Delta T \bullet \frac{\Delta z \cdot \Delta\varphi}{r}$	Voltage-dependent current source
2	$\Delta\varphi \bullet \Delta z\left[\frac{T_{j+\frac{\Delta r}{2}} - T_j}{\Delta r_f/2} - \frac{T_j - T_{j-\frac{\Delta r}{2}}}{\Delta r_f/2}\right]$	$R_{r,f} = \frac{\Delta r_f}{2\Delta z \cdot \Delta\varphi}$	Resistor
3	$\frac{\Delta r \bullet \Delta z}{r^2} \cdot \left[\frac{T_{\varphi+\frac{\Delta\varphi}{2}} - T_\varphi}{\Delta\varphi/2} - \frac{T_\varphi - T_{\varphi-\frac{\Delta\varphi}{2}}}{\Delta\varphi/2}\right]$	$R_{\varphi,f} = \frac{r^2 \cdot \Delta\varphi}{2\Delta r_f \cdot \Delta z}$	Resistor
4	$\Delta r \bullet \Delta\varphi\left[\frac{T_{i+\frac{\Delta z}{2}} - T_i}{\Delta z/2} - \frac{T_i - T_{i-\frac{\Delta z}{2}}}{\Delta z/2}\right]$	$R_{z,f} = \frac{\Delta z}{2\Delta r_f \cdot \Delta\varphi}$	Resistor
5	$\left(\frac{-\rho \cdot c \cdot u}{k}\right)_f \cdot \left[T_{i+\frac{\Delta z}{2}} - T_{i-\frac{\Delta z}{2}}\right]$	$G_{z,f} = -\Delta T \cdot \Delta r_f \cdot \Delta\varphi \bullet \left(\frac{\rho \cdot c \cdot u}{k}\right)_f$	Voltage-dependent current source

Table 4.
Formulae and analogous electrical devices yielded from fluid Eq. (13).

Term	Discretisation of equations	Thermal-electrical analogy	Type of device
1	$\Delta y \bullet \Delta z\left[\frac{T_{k+\frac{\Delta x}{2}} - T_k}{\Delta x/2} - \frac{T_k - T_{k-\frac{\Delta x}{2}}}{\Delta x/2}\right]$	$R_{x,a} = \frac{\Delta x}{2\Delta y \cdot \Delta z}$	Resistor
2	$\Delta x \bullet \Delta z\left[\frac{T_{l+\frac{\Delta y}{2}} - T_l}{\Delta y/2} - \frac{T_l - T_{l-\frac{\Delta y}{2}}}{\Delta y/2}\right]$	$R_{y,a} = \frac{\Delta y}{2\Delta x \cdot \Delta z}$	Resistor
3	$\Delta x \bullet \Delta y \frac{\left[T_{i+\frac{\Delta z}{2}} - T_i - T_i - T_{i-\frac{\Delta z}{2}}\right]}{(\Delta z/2)}$	$R_{z,a} = \frac{\Delta z}{2\Delta x \cdot \Delta y}$	Resistor

Table 5.
Formulae and analogous electrical devices yielded from fin Eq. (14).

only three terms, one for each Cartesian direction. **Tables 3–5** summarise the devices used in the elementary circuits.

A.4 Building the network model or whole equivalent circuit

Once the elementary circuits of the different parts of the system have been built, the equivalent circuit of each region is assembled by adding as many cells as needed according to the discretisation carried out; then the circuits of the regions are joined according to the topology.

Finally, the problem boundary conditions, Eqs. (4)–(11), must be implemented. In this case, the most relevant are voltage sources for Eqs. (4), (7) and (9), very large resistance resistors are used to emulate the adiabatic conditions, Eqs. (4), (5), (8), (10) and (11) or just electrical continuity, Eqs. (6) and (7bis).

A.5 Writing the code

One of the advantages of NSM is the use of well-known and reliable software for the analysis of electronic and electrical circuits (NGSpice, PSpice or others), which are also very easy to program. Naturally, for the programming of the elementary (and complete) circuit, we direct the reader to the corresponding websites [40] and literature [41], where the documentation can be found.

B. Simulating the electric circuits

The whole electric circuit (network model) is solved using appropriate software, such as NGSpice, PSpice or others.

B.1 Obtaining results

The results obtained must be interpreted taking into account that in the thermo-electric analogy, voltages and currents are analogous to temperature and heat flow, respectively. Both temperature profile along the pipe-fin and temperature maps of the cross section are plotted using suitable software from the voltages at the appropriate points along the pipeline or the selected cross section using the software.

Formulae and analogous electrical devices yielded from Eqs. (12)–(14).

Author details

Mariano Alarcón[1]*, Manuel Seco-Nicolás[1], Juan Pedro Luna-Abad[2] and Alfonso P. Ramallo-González[3]

1 Electromagnetism and Electronics Department, International Campus of Excellence in the European context (CEIR) Campus Mare Nostrum, University of Murcia, Murcia, Spain

2 Thermal and Fluid Engineering Department, International Campus of Excellence in the European context (CEIR) Campus Mare Nostrum, Technical University of Cartagena, Cartagena, Spain

3 Information and Communication Engineering Department, International Campus of Excellence in the European context (CEIR) Campus Mare Nostrum, University of Murcia, Murcia, Spain

*Address all correspondence to: mariano@um.es

References

[1] Zambolin E, Del Col D. Experimental analysis of thermal performance of flat plate and evacuated tube solar collectors in stationary standard and daily conditions. Solar Energy. 2010;**84**(8): 1382-1396

[2] Balaji K, Iniyan S, Goic R. Thermal performance of solar water heater using velocity enhance. Renewable Energy. 2018

[3] Balaji K, Idrish Khan A, Ganesh Kumar P, Iniyan S, Goic R. Experimental analysis on free convection effect using two different thermal performance enhancers in absorber tube of a forced circulation flat plate solar water heater. Solar Energy. 2019:445-454

[4] El-Maghlany WM, Minea AA. Novel empirical correlation for ionanofluid PEC inside tube subjected to heat flux with application to solar energy. Journal of Thermal Analysis and Calorimetry. 2019;**135**(2):1161-1170

[5] Mateo M, Echarri R, Samsón I. Thermal analysis and experimental validation of parabolic trough collector for solar adsorption refrigerator. Energy and Power Engineering. 2017;**09**(11): 687-702

[6] Arani AAA, Kazemi M. Analysis of fluid flow and heat transfer of nanofluid inside triangular enclosure equipped with rotational obstacle. Journal of Mechanical Science and Technology. 2019;**33**:4917-4929

[7] Kalteh M, Abbassi A, Saffar-Avval M, Harting J. Eulerian–Eulerian two-phase numerical simulation of nanofluid laminar forced convection in a microchannel. International Journal of Heat and Fluid Flow. 2011;**32**(1): 107-116

[8] Bejan A. General criterion for rating heat-exchanger performance. International Journal of Heat and Mass Transfer. 1978;**21**(5):655-658

[9] International Energy Agency (IEA). Natural gas information: Overview, IEA, Paris, 2021. 2021. Available from: https://www.iea.org/reports/natural-gas-information-overview

[10] Russell M. The Nord stream 2 pipeline economic, environmental and geopolitical issues. 2021

[11] Quintino FM, Nascimento N, Fernandes EC. Aspects of hydrogen and biomethane introduction in natural gas infrastructure and equipment. Hydrogen. 2021;**2**(3):301-318

[12] Barrett CB et al. A Profuse Pipeline of Promising Options. Palgrave Macmillan, Cham; 2022. pp. 73-158

[13] Petukhov B, Polyakov AF. Heat Transfer in Turbulent Mixed Convection. Berlin Heidelberg: Springer; 1988

[14] Bejan A. Convection Heat Transfer. Fourth ed. Hoboken, New Jersey: John Wiley & Sons, Inc.; 2013

[15] Çengel YA. Heat and mass transfer. Third ed. México: McGraw-Hill; 2007

[16] Wang J, Li J, Jackson JD. A study of the influence of buoyancy on turbulent flow in a vertical plane passage. International Journal of Heat and Fluid Flow. 2004;**25**(3):420-430

[17] Kim WS, Jackson JD, He S, Li J. Performance of a variety of low Reynolds number turbulence models applied to mixed convection heat

transfer to air flowing upwards in a vertical tube. ARCHIVE Proceedings of the Institution of Mechanical Engineers Part C Journal of Mechanical Engineering Science 1989-1996. 2004; **218**(11):1361-1372

[18] Graetz L. Ueber die Wärmeleitungsfähigkeit von Flüssigkeiten. Annalen der Physik. 1882; **254**:79-94

[19] Papoutsakis E, Ramkrishna D, Lim HC. The extended Graetz problem with prescribed wall flux. AICHE Journal. 1980;**26**(5):779-787

[20] Bilir Ş. Numerical solution of graetz problem with axial conduction. Numerical Heat Transfer, Part A: Applications. 1992;**21**(4):493-500

[21] Bılır Ş. Laminar flow heat transfer in pipes including two-dimensional wall and fluid axial conduction. International Journal of Heat and Mass Transfer. 1995; **38**(9):1619-1625

[22] Zueco J, Alhama F, González Fernández CF. Analysis of laminar forced convection with network simulation in thermal entrance region of ducts. International Journal of Thermal Sciences. 2004;**43**(5):443-451

[23] Cossali GE. Analytical solution of Graetz problem in pipe flow with periodic inlet temperature variation. International Journal of Heat and Mass Transfer. 2009;**52**(13–14): 3396-3401

[24] Ateş A, Darıcı S, Bilir Ş. Unsteady conjugated heat transfer in thick walled pipes involving two-dimensional wall and axial fluid conduction with uniform heat flux boundary condition. International Journal of Heat and Mass Transfer. 2010;**53**(23–24):5058-5064

[25] Altun AH, Bilir Ş, Ateş A. Transient conjugated heat transfer in thermally developing laminar flow in thick walled pipes and minipipes with time periodically varying wall temperature boundary condition. International Journal of Heat and Mass Transfer. 2016; **92**:643-657

[26] Seco-Nicolás M, Alarcón M, Alhama F. Thermal behavior of fluid within pipes based on discriminated dimensional analysis. An improved approach to universal curves. Applied Thermal Engineering. 2018;**131**:54-69

[27] Arutyunyan RV. Three-dimensional simulation of heat and mass transfer processes during shf heating of oil flow in a pipeline. Journal of Engineering Physics and Thermophysics. 2021;**94**(3): 745-752

[28] Stepanov AV, Egorova GN. Simulation of heat transfer of heating-system and water pipelines under northern conditions. Journal of Engineering Physics and Thermophysics. 2016;**89**(5):1284-1288

[29] Na T, Baodong C, Limin H, Qisheng C, Qi W. Research on the overall heat transfer coefficients of crude oil pipeline and product pipeline laid in one ditch. Petroleum Science and Technology. 2012;**30**(3):247-255

[30] Li Y, Liu Y. A mathematical model of heat transfer in problems of pipeline plugging agent freezing induced by liquid nitrogen. Fluid Dynamics & Materials Processing. 2022; **18**(3):775-788

[31] Dehdarinejad M, Behbahani-Nejad M, Hajidavalloo E. Experimental and numerical investigation of heat loss in transition modes of buried hot fuel oil pipeline. Proceedings of the Institution of Mechanical Engineers Part C Journal

of Mechanical Engineering Science 1989-1996. 2021;**235**(22):6567-6576

[32] Bathe K-J. Finite element procedures. New Jersey: Prentice Hall; 1996

[33] Bèg ÒÀ, Takhar HS, Soundalgekar VM. Thermoconvective flow in a saturated, isotropic, homogeneous porous medium using Brinkman's model: numerical study. International Journal of Numerical Methods for Heat and Fluid Flow. 1998; **8**(5):559-589

[34] Campo A, Arıcı M. Semi-analytical solution of the graetz problem with uniform wall heat flux utilizing the transversal method of lines. Journal of Heat Transfer. 2020;**142**(4)

[35] González-Fernández CF. Heat transfer and the network simulation method. In: Horno J, editor. Network Simulation Method. Research Signpost: Kerala (India); 2002

[36] González-Fernández CF, Alhama F, Alarcón M, López-Sánchez JF. Digital simulation of transient heat conduction with polynomial variable thermal conductivity and specific heat. Computer Physics Communications. 1998;**111**(1–3):53-58

[37] Luna-Abad JP, Alhama F, Campo A. The use of relative inverse thermal admittance for the characterization and optimization of fin-wall assemblies. Thermal Science. 2016;**2016**:151-160

[38] Illán F, Alarcón M. Optimization of annular cylindrical and spherical fins in an internal combustion engine under realistic conditions. Journal of Thermal Science and Engineering Applications. 2010;**2**(4):041002 (7 pages)

[39] Alarcón M, Alhama F, González-Fernández CF. Transient conduction in a fin-wall assembly with harmonic excitation-network thermal admittance. Heat Transfer Engineering. 2002;**23**(2): 31-43

[40] ngspice. Open source spice simulator. Available from: https:// ngspice.sourceforge.io/

[41] Corporation M. PSpice. 6th ed. California: Irvine; 1994

[42] Seco-Nicolás M, Alarcón M, Luna-Abad JP. 3D numerical simulation of laminar forced-convection flow subjected to asymmetric thermal conditions. An application to solar thermal collectors. Solar Energy. 2021; **220**:230–245

[43] Maré T, Galanis N, Voicu I, Miriel J. Experimental analysis of mixed convection in inclined tubes. Applied Thermal Engineering. 2006;**26**(14–15): 1677-1683

[44] Suryanarayana NV. Forced convection - internal flows. In: Kreith F, editor. The CRC Handbook of Thermal Engineering. Boca Ratón, London, New York, Washington, D.C.: CRC Press LLC; 2000. pp. 3-47-3-55

[45] Incropera FP, DeWitt DP, Bergman TL, Lavine AS. Fundamentals of Heat and Mass Transfer. 6th ed. Hoboken, New Jersey: John Wiley & Sons, Inc.; 2007

[46] Jiji LM. Heat Convection. Springer Science Business Media; 2009

[47] Shah RK, London AL. Laminar Flow Forced Convection in Ducts. New York: Academic Press; 1978

Chapter 8

Review on the Stability of the Nanofluids

Sumit Kumar Singh

Abstract

Both mono and hybrid nanofluids, the engineered colloidal mixture made of the base fluid and nanoparticles, have shown many interesting properties and become a high potential next-generation heat transfer fluid in various engineering applications. The present review focuses on improving the stability of the nanofluids. For this, the present review briefly summarizes the impact of nanofluid preparation on the stability of various nanofluids and described in the following classification; (a) Nanofluid constituent, (b) Nanomaterial synthesis, and (c) Nanofluid synthesis techniques which are well-grouped and thoroughly discussed. Physical mechanisms for heat transfer enhancement using nanofluids are explored as well. Most of the studies reveal that there are significant improvements in the stability of the nanofluids. Hence, there is an excellent opportunity to use stabled nanofluids in various engineering applications. Finally, some useful recommendations are also provided.

Keywords: nanofluid, stability, nanomaterial synthesis, surfactant

1. Introduction

Nanofluids are engineered by dispersing nanoparticles, having average sizes below 100 nm, in conventional heat transfer fluids. Proper and stable dispersion of even a negligible fraction of particles in nanofluids can offer significant enhancement in the heat transfer properties. Various types of nanoparticles like metals, metal oxides, alloys, allotropes of carbon, ceramics, phase change materials, and metal carbides are being used for preparing nanofluids. In addition to nanofluids, hybrid nanofluids have also gained attention recently due to significant improvement in heat transfer characteristics and stability may be caused by the synergistic effect of hybridization. Heat exchangers that use tubes or pipes often have a circular, rectangular, or elliptical cross-section and are easier to design. Tubular heat exchangers are fairly prevalent in pipeline engineering applications. These heat exchangers might be built to handle fluids under high pressure or to handle pressure differentials between cold and hot fluids. Double-pipe and shell-tube heat exchangers are additional categories that apply to these heat exchangers. Modifying the fluids' characteristics can also increase the heat exchange rate of a heat exchanger. Due to the fact that stable nanofluids have significantly improved heat transfer characteristics, particularly in terms of thermal conductivity, slip mechanisms, and the nanofin effect, they may be employed in tubular heat exchangers to increase energy efficiency. For preparing mono or

IntechOpen

hybrid nanofluids, the two-step method is generally used where firstly different nanoparticles or nanocomposites are prepared. Then they are mixed in the base fluid through magnetic or mechanical stirring. After that, the solution is sonicated and then characterized using different techniques to assure the proper (homogeneous) mixing and stability of the hybrid nanofluids. Both mono and hybrid nanofluids are thus prepared to provide improved heat transfer characteristics due to an increase in thermal conductivity, Brownian motion, proper dispersion, agglomeration, solid/ liquid interface layering, thermophoresis, the improved thermal network between the solid nanoparticle and fluid molecules, nanofin and nanoporous effects at the heat transfer surface. The reason behind this improvement can be summarized as: (i) More heat transfer surface between nanoparticles and fluid, (ii) Collision between the nanoparticles, (iii) Increment in the thermal conductivity due to the interactive effect of different nanoparticles, and (iv) Proper dispersion of the nanoparticles in the base fluid, creating micro turbulences. Therefore, in hybrid nanofluids, both nanoparticles compromise their properties and provide better thermo-physical, chemical, and rheological properties within the low cost that makes it preferable over nanofluids for different applications. Stability is the main key factor for the performance of nanofluids in various engineering applications. All the thermo-physical properties of nanofluid are dependent on its stability. The unstability of nanofluid can inhibit its performance in several applications such as heat exchangers, chemical industry applications, enhanced oil recovery etc. The unstability of nanofluid is caused due to the propensity of nanoparticles to form a cluster in the fluid. The nanofluids may be broadly categorized into three groups based on the nanoparticle composition, namely: (i) mono-nanofluids (made from one type of nanoparticles), (ii) hybrid nanofluids containing different nanoparticles, and (iii) hybrid nanofluids consisting one solid covered by a layer of another solid (composite nanoparticles).

The current review emphasizes the impact of nanofluid preparation on the stability of various nanofluids and is described in the following classification; (a) Nanofluid constituent, (b) Nanomaterial synthesis, and (c) Nanofluid synthesis techniques.

2. Literature review

The available literature on the preparation, characterization, and stability of mono/ hybrid nanofluids are discussed in three sections. In all section, it summarizes the impact of nanofluid preparation on the stability of various nanofluids and described in the following classification; (a) Nanofluid constituent, (b) Nanomaterial synthesis, and (c) Nanofluid synthesis techniques. Exclusive reviews on the heat transfer, pressure drop characteristics, and energy performance of both double-tube and shell-tube heat exchangers using nanofluids are presented in the third and fourth sections.

2.1 Impact of nanofluid preparation

There are two main approaches to synthesize nanofluids: the single-step method and the two-step method. In the one-step method, nanofluid is prepared directly by dispersing nanoparticles in the base fluid without the requirement of numerous steps such as particle drying, storage, etc. Using this method, the stability of nanofluid exhibits most superior compared to the two-step method. But this technique is not beneficial for large scale because of its high production cost. Therefore, the two-step method is the more effective and generally common method of nanofluid preparation. The foremost disadvantage of this process is the control of particle agglomeration tendency. The common application of wide ultrasonication and stirring is the most frequently used method to control agglomeration. Several forces such as Van der Waal attractive force, gravitational force, buoyancy force, and electrostatic repulsive force are acted which lead to destabilization and form sediments. The Van der Waal attractive force and gravitational force work against the stability of any colloidal suspension. Stability is the main key factor for the performance of nanofluids in various engineering applications. All the thermo-physical properties of nanofluid are dependent on its stability. The unstability of nanofluid can inhibit its performance in several applications such as heat exchangers, chemical industry applications, enhanced oil recovery etc. The unstability of nanofluid is caused due to the propensity of nanoparticles to form a cluster in the fluid. For considering a stable nanofluid, agglomeration propensity has to be removed. Some stability evaluation methods are used in literature i.e., sedimentation and centrifugation method, zeta potential measurement, spectral absorbance and transmittance measurement, and dynamic light scattering. Numerous efforts have been made to prepare long-time stable and homogenous nanofluids using various techniques. The current review emphasizes the impact of nanofluid preparation on the stability of various nanofluids and is described in the following classification; (a) Nanofluid constituent, (b) Nanomaterial synthesis, and (c) Nanofluid synthesis techniques.

2.1.1 Nanofluid constituent

2.1.1.1 Nanomaterial type

There are several types of nanofluids: metallic nanofluids (Al, Ag, Cu, Fe, Au), metal oxide nanofluids (Al_2O_3, CuO, Fe_3O_4, SiO_2, TiO_2, ZnO, etc.), and non-metallic nanofluids (SiC, TiC, graphite, diamond, SWCNT/MWCNT, graphene, etc.). Several studies on the impact of the nanofluid constituents on its stability are shown in **Table 1**. Xu et al. [4] prepared hybrid nanofluids with nanoparticles of different masses added with a small amount of SDBS and PEG into DW and observed that 25% Al_2O_3 + 75% TiO_2 hybrid nanofluid shows good suspension stability. The zeta potential value for the 25% Al_2O_3 + 75% TiO_2 hybrid nanofluid is found 42.6 mV indicating high stability. Zeta potential means electrostatic repulsion force between nanoparticles and base fluid. High repulsion force indicates high stability of nanofluid, whereby 30mV is generally considered as a benchmark for a stable nanofluid and excellent nanofluid stability may exceed 60 mV. Some studies investigated the impact of functionalizing the nanoparticles surface which reduces aggregation and improves dispersion. Said et al. [5] studied the stability of Carbon nanofiber (CNF), Functionalized Carbon nanofiber (F-CNF), Reduced graphene oxide (rGO), and F-CNF/rGO nanofluids. The results indicated that hybrid (FCNF/rGO) nanofluid shows higher stability than as compared to CNF, F-CNF, and rGO nanofluids. Also, the sample of CNF almost completely sedimented on 2nd day as shown in the **Figure 1**. It is due to the low charge density on the surface of the CNF nanoparticle which leads to the tendency of agglomeration. Said et al. [15] used acid treatment of CNF to examine the stability. The zeta potential of 0.02 vol. % F-CNF nanofluid was −42.9 and − 41.8 mV after 2 and 90 days which indicates that the stability was improved while the zeta potential of CNF was −16.3 and − 15.5 mV, indicating a relatively unstable dispersion. One way to achieve long-term stability is to adjust

Year	Author	Base fluid	Nanomaterial	Surfactant	Important results	Remarks
2020	Xian et al. [1]	DW-EG	COOH-GnP, TiO_2	SDC CTAB SDBS	CTAB shows highest degree of stability	Use of ionic surfactant results in higher stability
2019	Almanassra et al. [2]	Water	CNT	GA PVP SDS	GA and PVP show stablility for more than 6 months	GA can be a promising surfactant for stabilizing the CNT/W nanofluids
2020	Cacua et al. [3]	DI water	Al_2O_3	SDBS CTAB	Al_2O_3 nanofluid with SDBS at 1 CMC and CTAB at 0.5 CMC selects as the most stable and unstable nanofluids, respectively.	Anionic SDBS allows to have high repulsive forces between nanoparticles
2020	Xu et al. [4]	DW	Al_2O_3, TiO_2	SDBS PEG	25%Al_2O_3 + 75%TiO_2 shows better stability than other mono/hybrid nanofluid	Zeta potential values for the hybrid nanofluid and TiO_2/W nanofluids are 42.6 mV and 40.8 mV, respectively.
2020	Said et al. [5]	DW	CNF, F-CNF, rGO, F-CNF/rGO		Hybrid (FCNF/rGO) nanofluid shows a higher stability	CNF nanofluids can be better for high-temperature applications Hybrid nanofluid (FCNF/rGO) can be ideal working fluid for lower temperature
2020	Kazemi et al. [6]	Water	SiO_2, Graphene	CMC	G/water achieves better stability at higher pH values	Adjusting the nanofluid pH may lead to better stability
2019	Ouikhalfan et al. [7]	DW	TiO_2	CTAB, SDS	Using CTAB surfactant, nanofluid shows better stability	
2019	Siddiqui et al. [8]	DI Water	Cu, Al_2O_3, Cu-Al_2O_3		Al2O3 nanofluid shows excellent stability between 0 and 6 h. Cu nanofluid shows very low dispersion stability at time t = 1 h	Hybrid nanofluid with optimum mixing ratio exhibits relatively better stability
2018	Akhgar and Toghraie [9]	Water	TiO_2, MWCNT	CTAB	TiO_2 / Water with higher pH had more stability MWCNT particles are not dispersed in water and are not stable in any of pH	Low amount of surfactant gives better nanofluid stability.

Year	Author	Base fluid	Nanomaterial	Surfactant	Important results	Remarks
2018	Choi et al. [10]	DI Water	MWCNT	SDBS, CTAB, SDS, TX-100	For short-term time frames (3 h) nanofluids manufactured with SDBS, CTAB, and TX-100 show better stability For long-term time frames (1 month), the SDBS and TX-100 nanofluids have the highest suspension stability	TX-100, CTAB, and SDS are not suitable surfactants for nanofluids operating from 10–85°C
2018	Das et al. [11]	DW	TiO_2 (Anatase)	SDBS, CTAB, SDS, Acetic acid	SDS and CTAB results show excellent stabilization (stable for exceeding 12 h and 24 h)	
2018	Gao et al. [12]	DW, EG, EG/DW	GNP		Stability of GPN/DW nanofluid is worse than that of GNP/ EG.	
2018	Kuang et al. [13]	Brine	SiO_x Al_2O_3 TiO2	OA, PAA, Cationic, Anionic, Nonionic	SiO_x nanofluids exhibit stable in all cases Al_2O_3 + PAA and Al_2O_3+ cationic surfactant show the most stability	Among all surfactants, TiO_2 + PAA show the most stability
2018	Muthoka et al. [14]	PCM-DI Water	MgO, MWCNT	SDS	Stability of MgO and 24 wt% base fluid without surfactant showed poor stability after only 24 h Fuctionalized MWCNT nanofluid showed no separation after 24 h	Stability of the nanofluid at low temperatures is increased by the use of surfactant
2018	Said et al. [15]	DI water	CNF, F-CNF		F-CNF-based nanouids exhibit superior stability in water for 90 days	Acid treatment of CNF at relatively low temperatures resulted in stable suspension
2019	Alawi et al. [16]	DI Water	PEG-GnP, PEG-TGr, Al_2O_3, SiO_2		PEG-GnP shows higher dispersibility and the relative concentration of PEG-GnP based water reported very low sedimentation	
2019	Akbari and Saidi [17]	DW	TiO_2 GnP		TiO_2/DW nanofluid shows the good stability	Graphene is an inherently hydrophobic material and the stability of graphene/water nanofluid is not favorable

Year	Author	Base fluid	Nanomaterial	Surfactant	Important results	Remarks
2019	Cacua et al. [18]	DI	Al_2O_3	SDBS CTAB	Al_2O_3-SDBS exhibits lower rate of sedimentation compared to Al_2O_3 and Al2O3-CTAB.	
2019	Etedali et al. [19]	DI	SiO_2	Ps20, CTAB, SLS	Nanofluid with all surfactant show excellent stability	Maximum surface charge for the nanofluids with SLS, CTAB, and Ps20 surfactants are −87.4, 74.2, and − 97.9, respectively
2020	Giwa et al. [20]	DW, EG-DW	Al_2O_3-Fe_2O_3	SDS, NaDBS	DW-based Al_2O_3-Fe_2O_3 exhibited more stable than the EG-DW based Al_2O_3-Fe_2O_3	Absorbance value of the DW-based Al_2O_3-Fe_2O_3 displayed better horizontal straight lines than those of the EG–DW Al_2O_3-Fe_2O_3
2020	Kazemi et al. [21]	DW	G, SiO_2, G-SiO_2	CMC	SiO_2/Water nanofluid shows excellent stability at all pH value G/Water sustainability is poor in lower pH values	Due to better stability in higher pH values, the CMC surfactant can be used to increase pH by creating a negative charge surface for graphene nanoparticles and developing functional groups
2020	Boroomandpour et al. [22]	Water-EG	MWCNT, ZNO, TiO_2 MWCNT-ZNO-TiO2	CTAB	All nanofluids have good stability 48 h after fabrication	Addition of CTAB surfactant lead to better stability

Table 1.
Synopsis of the investigations about the impact of the nanofluid constituents on its stability.

the nanofluid pH, away from the isoelectric point (IEP). Thus, IEP differs from one sample to another. These values were prepared in acidic and alkaline ranges using HCl and NaOH solutions and adjusted by pH meter. Kazemi et al. [6] used two different nanoparticles (GnP, SiO_2) with the same base fluid (water) as well as different pH values (3,6,9, and 12) to study the stability of the nanofluids. The results found that SiO_2/Water nanofluids have good stability at all pH values, especially for samples with pH >3 and GnP/water achieve better stability at higher pH values. Akhgar and Toghraie [9] examined the stability of water-based MWCNT and TiO_2 nanofluid at different pH (3, 6,9, and 12). The results observed that the nanofluid containing water/TiO_2 with pH = 9 had more stability than the rest of the samples. On the other hand, MWCNT particles are not dispersed in water and are not stable in any pH without any surfactant. Kazemi et al. [21] compared the stability of three types of nanofluids,

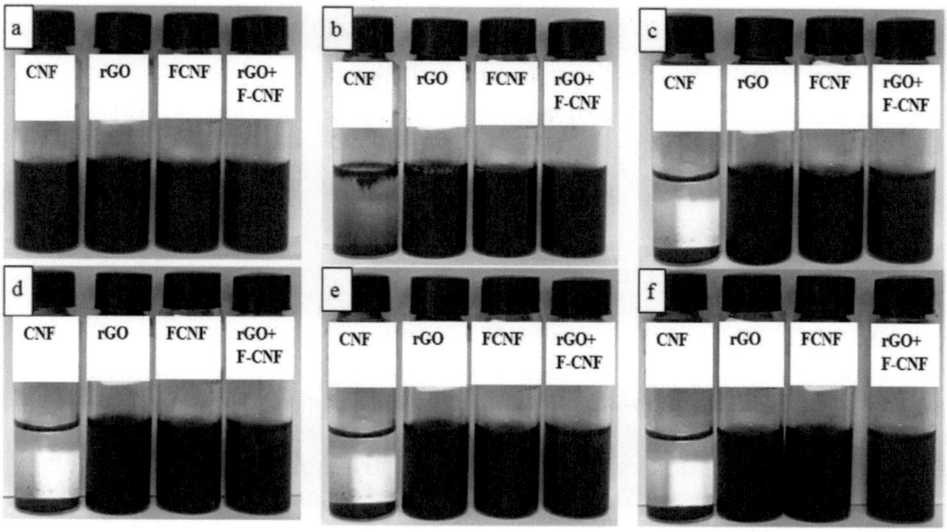

Figure 1.
Photographs of vials showing the stability of nanofluids for: (a) 1 st day, (b) 2nd day, (c) 30 days, (d) 45 days, (e) 60 days, and (f)180 days [5].

G/Water, SiO_2/Water, and G-SiO_2/Water and found that SiO_2/Water nanofluid shows excellent stability at all pH values while G/Water sustainability is poor in lower pH value. Due to better stability in higher pH values, the CMC surfactant can be used to increase pH by creating a negative charge surface for graphene nanoparticles and developing functional groups. Siddiqui et al. [8] performed a stability study with metal (Cu), oxide (Al_2O_3), and meta-oxide Cu- Al_2O_3) nanofluid containing the same base fluid (DI water). Al_2O_3 nanofluid exhibits better stability between 0 and 6 h followed by good stability with little particle settling between 6 and 240 h. Cu nanofluid shows poor dispersion stability after 1 hour of preparation. In case of hybrid nanofluid, nanofluid with optimum mixing ratio exhibits relatively better stability. Muthoka et al. [14] used PCM-DI water as the base fluid and two different nanoparticles (MgO and MWCNT) to examine the stability. They observed that the stability of MgO and 24 wt% base fluid without surfactant showed poor stability after only 24 h while the functionalized MWCNT nanofluid showed no separation after 24 h. Also, it was concluded that the stability of the nanofluid at low temperatures is increased by the use of surfactant. Alawi et al. [16] synthesized PEG-GnP, PEG-TGr, Al_2O_3, and SiO_2 water-based nanofluids. They observed the dispersion stabilities of carbon-based nanofluids and metallic oxides nanofluids for 30days, and the results showed the higher dispersibility of the PEG-GnP, PEG-TGr nanofluids in an aqueous media with very low sedimentation. Akbari and Saidi [17] observed TiO_2/DW nanofluid shows good stability as compared to GnP/DW nanofluid. Since graphene is an inherently hydrophobic material and the stability of graphene/water nanofluid is not favorable without any surfactants. Boroomandpour et al. [22] studied the stability of ternary hybrid nanofluids containing MWCNT-TiO_2-ZnO/DW-EG (80:20) as well as binary and mono nanofluids. They found that all nanofluids have good stability up to 48 h after fabrication and the addition of CTAB surfactant lead to better stability.

From the literature reviews on the preparation of nanofluids with different particles, it is found that the stability of water mono/hybrid nanofluid is strongly dependent on the particle shape and size. It is found that the propensity of aggregation is increased with the reduction in particle size and isoelectric point (pH value) decreases with the decrease in particle size. Therefore, the agglomeration process moves toward lesser pH value. The cylindrical-shaped particles sediment faster than spherical and platelet-shaped particles. High aspect ratio nanoparticles are more susceptible to agglomeration.

2.1.1.2 Surfactant type

Addition of different surfactants such as: Anionic (Sodium Dodecyl Sulfate (SDS), Sodium Dodecyl Benzene Sulfonate (SDBS)), Cationic (Cetyltrimethylammonum Bromide (CTAB)), Non-ionic (Span 80, Tween 20) and polymer (Polyvinyl Pyrrolidone (PVP), Poly Vinyl Alcohol (PVA), Gum Arabic (GA)) during nanofluid preparation is an additional way of controlling particle aggregation. A negatively charged suspension may be obtained by using anionic surfactants (SDC, SDS, and SDBS) while a cationic surfactant (CTAB) may contribute a positive charge. The augmentation in stability will achieve by the coating of surfactant on nanoparticles, which leads to a dominating electrostatic repulsion over the van der Waals force and thus prevent nanoparticles from agglomerating. Also, the stability of the nanofluid can be improved by decreasing the sedimentation velocity of the nanoparticles. According to Stokes law, the sedimentation velocity can be reduced by using nanoparticles with smaller diameters. However, when the nanoparticles diameter decreases, the surface energy will be increased which leads increase possibility of agglomeration. The best way to suppress the agglomeration without disturbing the sedimentation velocity is the usage of surfactants. While surfactant addition is an active way to improve the stability of the nanofluids but surfactants may lead to cause some problems. Surfactants may contaminate the heat transfer media. Surfactants may produce foams while heating and cooling are regular processes in heat exchange systems. Additionally, surfactant molecules attributed to the surfaces of nanoparticles might increase the thermal resistance between the nanoparticles and the base fluid, which may hinder the augmentation of the thermal conductivity.

Xian et al. [1] used three different surfactants, i.e., SDS, CTAB, and SDBS to stabilize the COOH-TiO_2 hybrid nanofluid. They observed that COOH-TiO_2 hybrid nanofluid with CTAB surfactant exhibited the best surfactant to stabilize this hybrid nanofluid. The visual inspection of sedimentation of nanofluids with different surfactants after 40 days is shown in **Figure 2**. Almanassra et al. [2] compared the effect of different types of surfactants on the stability of CNT/water nanofluids. They investigated with three types of surfactants namely, GA, PVP, and SDS and found that the nanofluids with GA as well as PVP surfactants were more stable for more than 6 months. Gum Arabic can be a promising surfactant for stabilizing the CNT in water-based nanofluids. Cacua et al. [3] found Al_2O_3 nanofluid with SDBS at 1 CMC and CTAB at 0.5 CMC were the most stable and unstable nanofluids, respectively. Anionic SDBS provides high repulsive forces between nanoparticles. Ouikhalfan et al. [7] prepared surface-modified TiO_2 nanofluid with two different surfactants (SDS and CTAB). The quick sedimentation was found in non-treated TiO_2 nanofluid after 24 hours of the preparation as shown in **Figure 3**. TiO_2 nanofluid with CTAB showed better stability up to several days while the nanofluid with SDS surfactant shows less but overall better dispersion compared to nanofluid with non-treated TiO_2. Choi et al. [10] studied the effect of various surfactants as well as the temperature on the stability of water-based MWCNT nanofluid. They prepared nanofluid with four different surfactants, i.e., SDBS, CTAB, SDS, and TX-100 between the

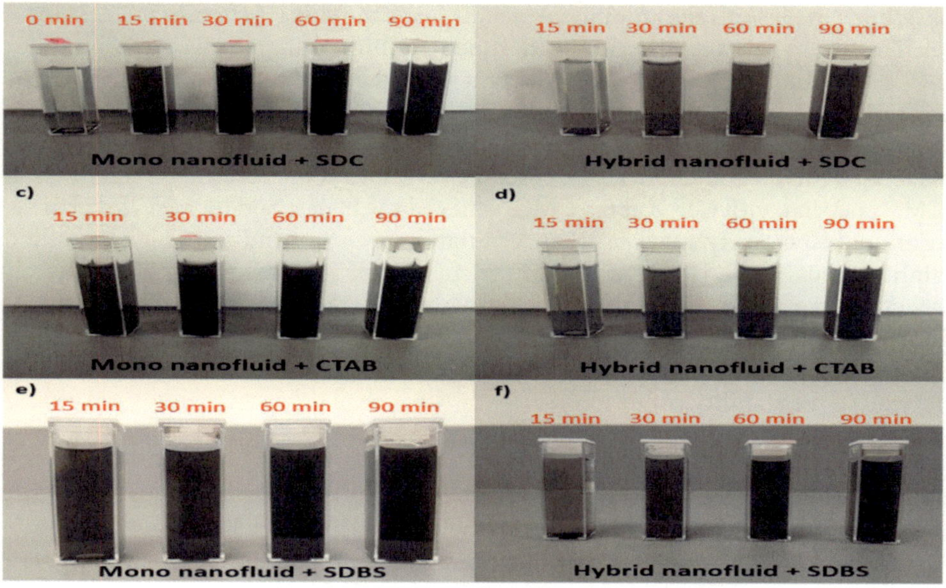

Figure 2.
Visual inspection of sedimentation of nanofluids with different surfactants and ultra-sonication time after 40 days. [1].

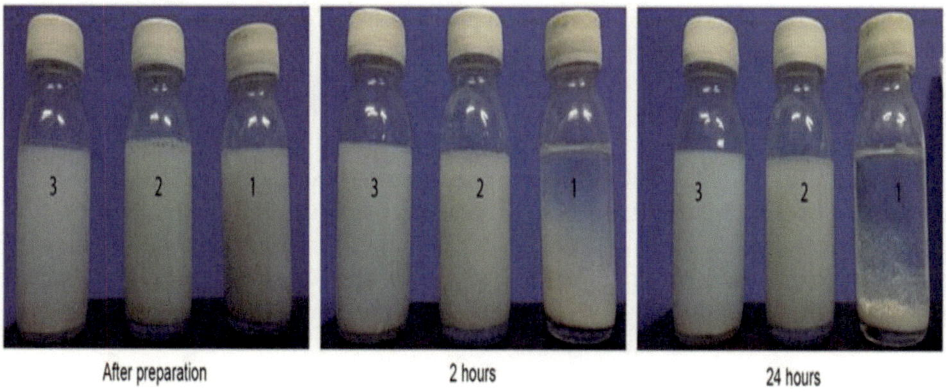

Figure 3.
Sediment photograph capturing of the nanofluid with (1) nontreated TiO_2, (2) CTAB-treated TiO_2 nanofluid, and (3) SDS-treated TiO_2 nanofluid [7].

temperatures 10°C–80°C. It was observed that for short-term time period (3 h), nanofluids prepared with SDBS, CTAB, and TX-100 show better stability while for long-term time period (1 month), the SDBS and TX-100 nanofluids have the highest suspension stability. On the account of temperature, TX-100, CTAB, and SDS are not suitable surfactants for nanofluids operating from 10 to 85°C. Das et al. [11] found TiO_2 (Anatase) with SDS and CTAB show excellent stabilization (stable for exceeding 12 h and 24 h) as compared with nanofluid with SDS and acetic acid surfactant. Kuang et al. [13] prepared nanofluids by dispersing three nanoparticles (i.e., SiOx, Al_2O_3, and TiO_2) and five different chemical agents i.e., oleic acid (OA), polyacrylic acid (PAA), a cationic, an anionic, and a nonionic surfactant) in base brine solutions.

Nanofluids made with the anionic surfactant made the surface slightly more water wet. The results revealed that SiO_x nanofluids exhibit stability in all cases while Al_2O_3 + PAA and Al_2O_3 + cationic surfactant show the most stability. In case of TiO_2 nanofluid, TiO_2 + PAA show the most stability among all surfactants. Cacua et al. [18] used UV–vis spectroscopy to examine the stability of Al_2O_3 with two different surfactants (SDBS and CTAB). The outcome reveals that the nanofluid with SDBS at 1 CMC and that with CTAB at 0.5 CMC achieved the lowest and highest absorbance variation, respectively. Low absorbance variation over time indicates high nanofluid stability. Etedali et al. [19] investigated the stability of SiO_2 nanofluids with different surfactants, i.e., SLS, CTAB, and Ps 20 through the Zeta potentials test. The results of the Zeta-potential test found that the maximum surface charge for the nanofluids with SLS, CTAB, and Ps20 surfactants were – 87.4, 74.2, and – 97.9, respectively, confirming the stability conditions.

2.1.1.3 Base fluid type

Gao et al. [12] prepared GNP nanofluid with three different base fluids namely, EG, DW, and EG/DW and reported that the stability of nanofluid with EG base fluid is better than that of DW-based nanofluid. **Figure 1** shows the visual observation of GNP nanofluid with different base fluids. Giwa et al. [20] used two-step method to prepare Al_2O_3-Fe_2O_3 hybrid nanofluid with two type of base fluid viz., DW and EG/DW. SDS and NaDBS were used as a surfactant. Using UV-visible spectrophotometer, they found DW-based Al_2O_3-Fe_2O_3 were relatively more stable than the EG–DW Al_2O_3-Fe_2O_3 hybrid nanofluid. The absorbance value of the DW-based Al_2O_3-Fe_2O_3 displayed better horizontal straight lines than those of the EG–DW Al_2O_3-Fe_2O_3.

2.2 Nanomaterial synthesis

Ding et al. [23] prepared the functionalized graphene (ESfG) by adding the graphite powder into the milling jar with steel balls of smaller diameter and the system was filled with SO_3 gas. After removing metallic impurities, the samples were then freeze-dried for 36 hours at –120°C to yield black powder as the final ESfG. The prepared ESfG was stable for several months in water. The sulfonic-acid groups can bond with carbon atoms at the edge of graphite which tends to enhance the stability of ESfG water-based nanofluids. Gul and Firdous [24] synthesized the graphene oxide nanosheet by the oxidation of graphite using the Hummers method as shown in **Figure 4**. In this method the graphite powder was mixed with $NaNO_3$, H_2SO_4, and $KMnO_4$ and stirred in an ice bath for about 30 min. Finally, the mixture was sonicated and added H_2O_2 and HCl to quench the reaction and get light yellow graphite oxide. The results found that the highly dispersible nature of GO in water which is fruitful for the preparation of GO nanofluid for multipurpose applications. Li et al. [25] introduced the β-cyclodextrin(β-CD) onto the surface of MWCNTs by a simple chemical synthesis method. It was found that the introduction of β-CD onto the surface of MWCNTs exhibited better stability of nanofluids. The possibility of aggregation between CD-CNTs is significantly decreased due to the Vander Waals force or steric interrupts between β-CD. Rahimi et al. [26] treated the hydrophilization of MWCNTs with different concentrated acids. They added the raw MWCNTs into the mixture of H_2SO_4 and HNO_3 and the mixture was refluxed for 3 hours. The acid-treated MWCNTs were obtained after washing with DI water and dried for 4 hours. Acid-treated MWCNTs suspensions display good stability in water. This is due to

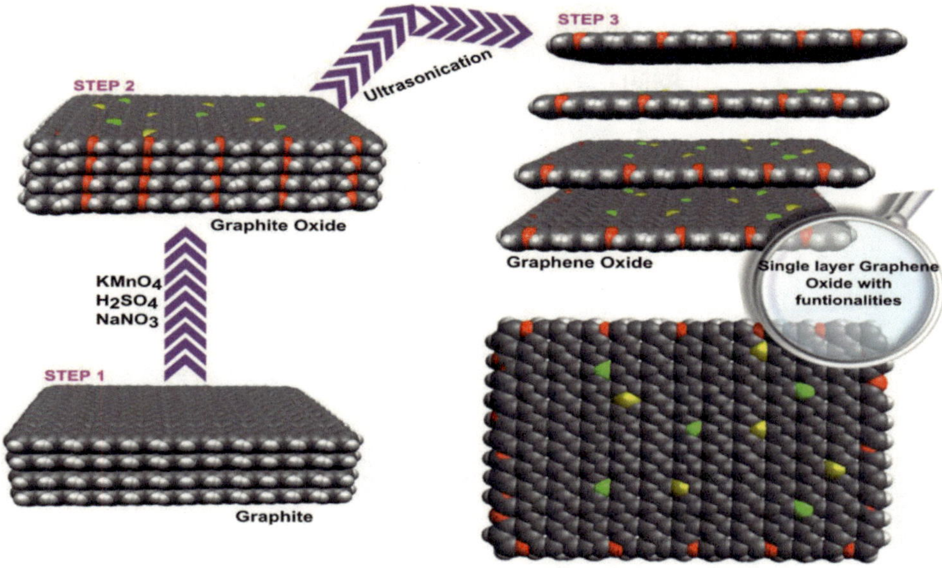

Figure 4.
Synthetic route of graphene oxide by hummers method [24].

the generation of hydroxyl groups on nanotube surfaces. Vozniakovskii et al. [27] synthesized a hybrid nanomaterial composed of nanodiamonds-multi-walled carbon nanotubes (DND-CNT) using a catalyst chemical vapor deposition (CCVD) method. The results showed that DND-CNT hybrid suspension was stable up to 100 hours while the initial DND began to precipitate after 1 hour. The stability of DND-CNT hybrid particles in water is explained by the opening of a previously closed surface covered with groups with a labile proton, which ensures the stability of the particles of the hybrid material in water.

2.3 Nanofluid synthesis technique

Numerous nanofluid stabilization techniques are used for reducing the cluster size of nanoparticles i.e. ultrasonic vibration and ball milling etc. The role of ultrasonication is to break the nanoparticle cluster and create a homogenous mixture. Ultrasonic vibration can be employed in two ways; (a) indirect method (ultrasonic bath), and (b) direct method (probe sonicator). Among these two methods, the probe sonicator offers better results in terms of breaking the particle cluster and lowering the average cluster size. Several studies on the impact of the nanofluid synthesis technique on its stability are shown in **Table 2**. Asadi et al. [28] used two-step method to prepare TiO_2-CuO hybrid nanofluid. They applied a magnetic stirrer for 1 hour in order to distribute the nanomaterial in the base fluid. Moreover, for breaking the clusters and uniformly distributing the nanoparticles in the base fluid, a probed ultrasonic device was applied for 1 hour. The DLS results ensured the nanoparticles exist in the base fluid, and the phenomenon of agglomeration did not happen. Chen et al. [29] investigated the impact of sonication time on the stability of the Al_2O_3/liquid paraffin nanofluid. They used two-step method with varying the magnetic stirrer time from 10 to 40 minutes and sonication time from 1 to 4 hours. It was found that nanofluids prepared using shorter sonication times show stability for a minimum of 1 month. When increase in sonication time, it breaks the bond between the nano additives and

Year	Author	Nanofluid	Synthesis technique	Important results	Remarks
2020	Asadi et al. [28]	CuO-TiO_2/Water	Two-step methods Magnetic stirrer time = 1 hour Sonication time = 1 hour	DLS results confirmed the nano-dimensions of the particles exist in the base fluid, and the phenomenon of agglomeration has not occurred.	Strong Van der Waals attraction forces among the particles, type and size of particles, and features of the base fluid may result in the occurrence of a phenomenon known as clustering of nanoparticles
2019	Chen et al. [29]	Al_2O_3/liquid paraffin	Two-step methods Magnetic stirrer time = 10–40 min Sonication time = 1–4 hour	Nanofluids prepared using shorter sonication times show stability for a minimum of 1 month	
2019	Asadi et al. [30]	MWCNT/Water	Two-step methods Magnetic stirrer time = 2 hour Sonication time = 10–80 min	Samples subjected to 10, 20, 40, and 60 min ultrasonication showed good stability	Applying ultrasonication longer than the optimum time leads to having agglomerated particles, which results in increasing the sedimentation rate
2018	Ranjbarzadeh et al. [31]	SiO_2/ Water	Two-step methods Magnetic stirrer time = 1 hour Sonication time = 1 hour	Long term stable nanofluid (for more than 6 months)	
2018	Aberoumand and Jafarimoghaddam [32]	Ag-WO_3/ Transformer oil	One-step method (EEW)	Results show very good stability for all of the test samples	
2018	Dalkılıç et al. [33]	CNT-SiO_2/DW	Two-step methods Sonication time = 3 hour	Sedimentation not observed up to 48 h	CNT particles show less stability in water without SiO_2 particles and surfactant
2018	Aberoumand et al. [34]	Ag/Water	One step method (EEW)	Zeta potential index confirmed the stability of the utilized nanofluid even after 1 year	EEW method may maintain the stability of nanofluid for a long time.
2018	Kakavandi and Akbari [35]	MWCNT-SiC/ Water-EG	Two-step methods Magnetic stirrer time = 1 hour Sonication time = 45 min	DLS test indicate acceptable stability of nanofluids	

Year	Author	Nanofluid	Synthesis technique	Important results	Remarks
2018	Keyvani et al. [36]	CeO_2/EG	Two-step methods Magnetic stirrer time = 2 hour Sonication time = 7 hour	Sedimentation of particles occurred after 2 weeks.	
2018	Liu et al. [37]	rGO/DI water	Two-step methods Sonication time = 30 min pH = 10	rGO nanofluids exhibited good stabilitiy for 10 day without the addition of other dispersants	UV–Vis intensity (at 270 nm) changes of rGO nanofluids for 10 days
2018	Ranjbarzadeh et al. [38]	GO-SiO_2/Water	Two-step methods Sonication time = 1 hour pH >7	Chemical reaction of the functional groups on surface of the particles to the base fluid promote stability of the nanofluid	
2018	Sharafeldin and Grof [39]	CeO_2/Water	Two-step methods Sonication time = 90 min	Higher concentration lead less stability	Mean value of zeta potential for 0.0666% volume fraction was −36.91 mV which indicate physical stability
2018	Zeng and Xuan [40]	MWCNT-SiO_2/Ag	Two-step methods Sonication time = 60 min	No precipitations found till 7 days, indicating their long-term dispersion stability	
2019	Gulzar et al. [41]	Al_2O_3-TiO_2 / Therminol-55	Two-step methods Magnetic stirrer time = 4 hour Sonication time = 2 hour	Value of zeta potential decreases with the increase in concentration which may cause agglomeration sufficiently after long time	The introduction of the functional group attached to oleic acid on the surface of both nanoparticles helps to reduce the attractive forces among them which prevents the agglomeration and increases the stability
2019	Alarifi et al. [42]	MWCNT-TiO_2/Oil	Two-step methods Magnetic stirrer time = 2 hour Sonication time = 1 hour	Stability of the prepared samples was observed over 14 days, and no sedimentation was observed	
2019	Akram et al. [43]	CGNP/DI Water	Two-step methods Sonication time = 1 hour	zeta potential values for the CGNP nanofluids are far from the isoelectric point (i.e., point of zero charge)	pH range (2.8–10.55) results in strong electric repulsion forces between the particles of CGNPs

Year	Author	Nanofluid	Synthesis technique	Important results	Remarks
2019	Sharafeldin and Grof [44]	WO_3/Water	Two-step methods Sonication time = 75 min	Mean zeta potential value for WO_3/water nanofluid was −43.12 and a little decrease in the values were observed along the period of 7 days.	
2019	Chen et al. [45]	Raw MWCNT, Acid treated MWCNT Milling treated MWCNT	Two-step methods, Stirred media mill technique	Excellent stable of fresh MWCNTs nanofluids is obtained by milling	
2019	Ali et al. [46]	Al/water	Two-Step Fabrication Approach, Controlled bath temperature two-step methods	30 °C nano suspensions showed better short- and long-term stability behavior than the conventionally fabricated nanofluids	
2019	Mahbubul et al. [47]	Al_2O_3/Water	Ultrasonication time = 1–5 hour	Higher sedimentation rate observed for the nanofluid prepared by low ultrasonication Increasing ultrasonication duration decreased the sedimentation rate.	Longer ultrasonication durations are necessary to avoid sedimentation if the nanofluids are stored for longer periods.
2019	Mahyari et al. [48]	GO-SiC/ Water-EG	Two-step methods Magnetic stirrer time = 1 hour Sonication time = 45 min	DLS test results with different patterns approved acceptable stability of the nanofluid.	
2017	Chen et al. [49]	Fe_3O_4-MWCNT/ Brine water	Two step methods Magnetic stirrer time = 30 min Sonication time = 2 hour	Magnetic MWCNTs nanofluids have high stability in 1000 ppm saline water, and long-term suspension stability also could be obtained,	When the solution salinity increased, the original colloidal structure destroyed by charge ion. Therefore, the salt-resisting surfactant was added to reinforce the double-layer repulsion and remained the system stability
2019	Okonkwo et al. [50]	Al_2O_3-Fe/Water	Two-step methods Sonication time = 8–9 hour pH = 2–12	Hybrid nanofluids are significantly more stable at pH values of 12 when compared at any other pH value.	High pH value favors the stability of the nanofluids

Year	Author	Nanofluid	Synthesis technique	Important results	Remarks
2019	Terueal et al. [51]	$MoSe_2$	LPE method Sonication frequency (kHz) = 80 and 130,	Nanofluid with 80 kHz and 130 kHz show the highest extinction coefficients values.	Higher extinction coefficient values means highest amount of nanomaterial in suspension
2020	Li et al. [52]	SiO_2-oleic acid/ liquid paraffin	Two-step methods Magnetic stirrer time = 30 min Sonication time = 1 hour	Large numbers of SiO_2 nanosized particles possesses maximum value for total count at values less than−40 mV	High stability of SiO_2 nano sized particles in was found liquid paraffin
2020	Geng et al. [53]	ZnO-MWCNT/ Oil	Two-step methods Magnetic stirrer time = 1 hour Sonication time = 1 hour	Results show that the nanoparticles are in nanoscale after the construction of nano-oil	
2020	Li et al. [54]	SiO_2/EG	Two-step methods	Critical voltage value is −56.28 mV and nanofluid is stable.	Greater the number of particles with a smaller diameter, the higher the probability of stability.

Table 2.
Synopsis of the investigations about the impact of the nanofluid synthesis technique on its stability.

the surfactant which leads to be unstable. Asadi et al. [30] varied the sonication time from 10 to 80 minutes to measure the stability of MWCNT/water nanofluid. They reported that after the 30th day, the samples subjected to 10, 20, 40, and 60 minutes of ultrasonication showed good stability while the samples subjected to longer time ultrasonication showed the amount of sedimentation leads to having agglomerated particles. Ranjbarzadeh et al. [31] used magnetic stirrer for 1 hour to mix the SiO_2 nanoparticles in the base fluid and then sonicated for 60 minutes. By visual observation, the result found that no sediments were formed after 6 months. Aberoumand and Jafarimoghaddam [32] prepared Ag-WO_3/Transformer oil nanofluid using the first step method. They applied Electrical Explosion Wire (EEW) to prepare the nanofluid. The Zeta potential of applied nanofluids in three different concentrations of 1%, 2%, and 4% was measured. The results indicate the excellent stability of applied hybrid nanofluids. Using the same EEW method, Aberoumand et al. [34] prepared Ag/water nanofluid and found that with EEW method, the nanofluid maintained their stability for a long time. Dalkılıç et al. [33] prepared CNT-SiO_2/DW using two-step methods and the mixture was sonicated for 3 hours. It was found that the sedimentation was not observed up to 48 hours. The raw CNT particles showed poor dispersion stability in the base fluid. SiO_2 particles support and increase the stability of CNTs particles in water. Tests showed that CNT particles exhibit less stability in water without SiO_2 particles and surfactants. Kakavandi and Akbari [35] used DLS test to examine the distribution of the MWCNT and SiC nanoparticles in the hybrid nanofluids. The results indicated acceptable stability of nanofluids. The hybrid nanofluid was magnetically stirred for 1 hour and then sonicated for 45 minutes. Keyvani et al. [36] used Ce_2O/EG nanofluid to examine the stability. The nanofluid was stirred and then exposed to ultrasonic waves for 2 and 7 hours, respectively. The sedimentation of particles was found after 2 weeks. Nanofluid with a higher concentration of particles, nanoparticles led to agglomerate; therefore, the stability of the nanofluid weakened. It was also reported that the stability of the prepared nanofluid with a lower volume fraction of nanoparticles was stable for a longer period of time compared to the nanofluids with a higher volume fraction [39].

Liu et al. [37] prepared rGO by the reduction of graphene oxide with L-ascorbic acid as a reductant in an aqueous solution. To prepare rGO, the graphene oxide solution was dispersed in DI water and ultrasonicated for 1 hour. NH_3-water was then added to control the pH to 10 with sonication for 30 minutes. L-ascorbic acid was added and the mixture was maintained at 95°C for 3 hours for the completion of the reaction. The rGO solution was filtered to obtain rGO on the filter paper. Finally, rGO nanofluids were prepared by sonicating the filtered powder in a certain amount of DI water. The whole process is shown in **Figure 5**. The rGO nanofluids exhibited good stability for 10 day without the addition of other dispersants. Ranjbarzadeh et al. [38] conducted a test to study of pH effects on the stability in acid and alkaline spectrums for GO-SiO_2/Water hybrid nanofluids. The results observed that the nanofluid, due to the presence of functional groups on the surface of its nanoparticles, shows acceptable stability in all spectrums; however, in the long term, nanofluids with $pH > 7$ showed better stability. Zeng and Xuan [40] sonicated the MWCNT-SiO_2/Ag binary nanofluids for 1 hour and reported that the stability of the binary nanofluid sustained the dispersion stability for 7 days. Gulzar et al. [41] dispersed hybrid nanopowder (Al_2O_3-TiO_2) in Therminol-55 oil and the mixture was subjected to high shear stirring at 2500 rpm using a magnetic stirrer for 4 hours. The mixture was then sonicated for 2 hours using a high energy probe sonicator. Oleic acid was used as a surfactant as of

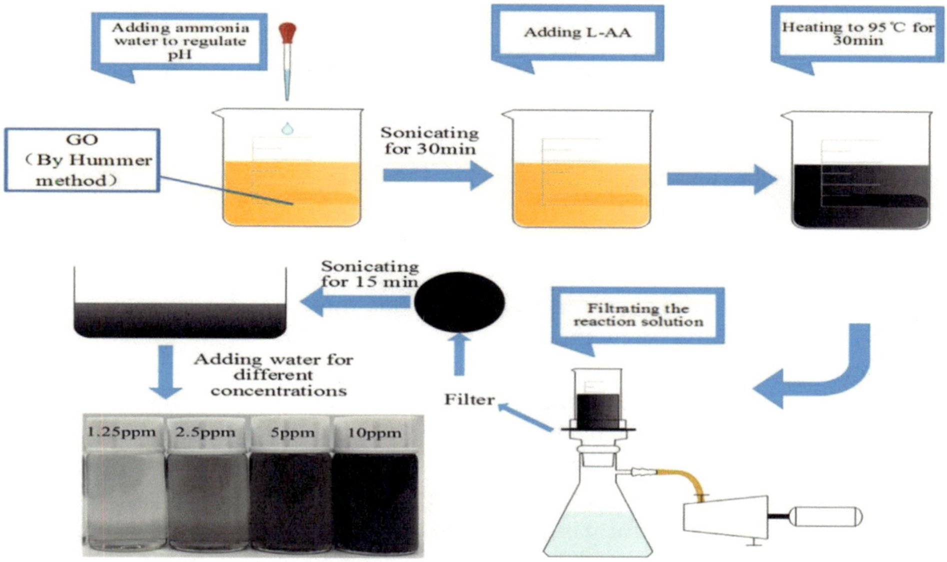

Figure 5.
Schematic graph of rGO nanofluids preparation.

better miscibility with Therminol-55 oil. They observed that the value of zeta potential declines with the rise in concentration which may cause agglomeration adequately after a long time. The surfactant which changes the surface charge and increases the repulsive forces between the nanoparticles also contributes to improved stability. Same way, Alarifi et al. [42] used magnetic stirring for 2 hours and sonicated for 1 hour to prepare a long-term stable MWCNT-TiO_2/oil nanofluid. The stability of the prepared hybrid nanofluid was observed over 14 days and no sedimentation was found. Akram et al. [43] checked the stability of CGNP–water nanofluid by zeta potential at different pH values. They prepared this nanofluid after the sonication for 60 minutes and observed that the CGNP nanofluid had high negative values (– 4.42 mV to –49.5 mV) within the pH variations from 1.84 to 10.55. The zeta potential values for the CGNP nanofluids are far from the isoelectric point (i.e., point of zero charges), which indicates that this pH range (2.8–10.55) results in strong electric repulsion forces. Sharafeldin and Grof [44] did sonication of WO_3/water nanofluid continuously for about 75 minutes to break the agglomeration between the nanoparticles which leads to well disperse the particles into water. The mean zeta potential value for WO_3/water nanofluid was –43.12 and a little decrease in the values was observed over the period of 7 days. MWCNT nanofluids suffer from low dispersion and short-time stability which inhibit their practical application. Chen et al. [45] used a novel method, i.e., a one-pot method by stirred media mill technique. In this method, raw MWCNTs nanoparticles were treated by ball milling to change their morphology, length, and specific surface first. After centrifuging, dry nanoparticles were purified by acid treatment to improve their dispersion in the solution. Thus, the resulting powder was dispersed again in base fluids by ultrasonication and meanwhile, surfactant was added to improve dispersion. The results showed that the milling-treated MWCNT nanofluid exhibited better stability as compared to raw MWCNT and Acid treated MWCNT nanofluid. Ali et al. [46] investigated the stability of dispersed Al nanoparticles in base fluid (water) prepared by the conventional and the controlled bath temperature two-step methods. The sonication process was taken the same for 4 hours in the range of 10–60°C. The results revealed that the sedimentation behavior of the nanofluids prepared through the controlled bath temperatures of less than 30°C was of dispersed sedimentation type, while those produced by the conventional method and the fixed temperatures of 30°C and higher were of flocculated sedimentation type. Furthermore, increasing the controlled sonication temperature led to an increase in the settling process of the sediments. Also, the rise in nanoparticle concentration was seen to reduce the variation in sedimentation height ratio between the fixed temperature samples. A comparison between the two preparation methods was shown that the 30°C nanofluids had better short- and long-term stability than the conventionally produced suspensions. Mahbubul et al. [47] varied the sonication time from 1 to 5 hours to study the effect of sonication time on the stability of the 0.5 vol% Al_2O_3 nanofluids. They observed that with low sonication time or no sonication, the sedimentation rate is higher. It can be concluded that longer ultrasonication reduces the sedimentation of nanoparticles and hence, increases the stability of nanofluids. Mahyari et al. [48] used probe-type ultrasonicator to achieve the stability of GO-SiC/water-EG hybrid nanofluid. DLS test results with different patterns approved acceptable stability of the nanofluid. Chen et al. [49] prepared the saline water based magnetic MWCNT nanofluids at different mass concentration from 0 to 0.04 wt% by two-step method. A mechanical stirrer was used at 500 rpm continuously for 30 minutes to mix nanoparticles and water and then the mixture was sonicated thoroughly for 2 hours. Magnetic MWCNTs nanofluids showed high stability in 1000 ppm saline water, and when the solution salinity increased, the original colloidal structure would be destroyed by charge ion. Therefore, the salt-resisting surfactant was added to reinforce the double-layer repulsion and remained the system stable. Okonkwo et al. [50] prepared the Al_2O_3-Fe/Water using two-step methods and measured the stability of nanofluid through the Zeta potential test. Hybrid nanofluid was found significantly more stable at pH values of 12 when compared at any other pH value. Terueal et al. [51] performed using the liquid phase exfoliation technique starting with bulk $MoSe_2$ to prepare stable nanofluids. Triton X-100 was used as a surfactant. The suspension underwent sonication in an ultrasound bath for 4 hours with two different frequencies: 80 kHz and 130 kHz. The samples were then centrifuged at 1000 rpm for 10 minutes and again at 4000 rpm for 10 minutes. The results showed that the nanofluid prepared with the frequency of 80 kHz and 130 kHz show the highest extinction coefficient values after 30 days. Higher extinction coefficient values mean the highest amount of nanomaterial in suspension. Li et al. [52] analyzed the stability of SiO_2-oleic acid/liquid paraffin nanofluid through the Zeta potential test. The nanofluid was prepared with two-step methods (magnetic stirrer for 30 minute and then sonicated for 1 hour). It was found that the large numbers of SiO_2 nano-sized particles possess maximum value for the total count at values less than −40 mV indicating high stability of SiO2 nano-sized particles in liquid paraffin. Geng et al. [53] used the DLS test to study the stability of ZnO-MWCNT/Oil nanofluid and found that the nanoparticles are in the nanoscale after the preparation of nano-oil. Li et al. [54] produced SiO_2/EG nanofluids by the two-step method with a mass fraction of 0.005–5%. The zeta potential value of the nanofluid was found −56.28 mV and claimed that the nanofluid is stable. Greater the number of particles with a smaller diameter, the higher the probability of stability. Nanofluid cluster formation may lead to the larger diameter of the nanoparticles. As the number of clusters increases, the fluid stability will decrease.

3. Conclusion

From the literature, it can be concluded that the stability of suspension of nanoparticles in the base fluid is improved when the nanofluid is synthesized by the one-step method as compared to the two-step method but the preparation of nanofluids by one-step method is difficult and expensive relative than two-step method. The literature also reveals that with low sonication time or no sonication, the sedimentation rate is higher. It can be concluded that longer ultrasonication reduces sedimentation of nanoparticles and hence, increases the stability of nanofluids. There are major tasks, which need to be focused on for selection of mono/hybrid nanofluids and their fabrication process, the stability of hybrid nanofluids. The stabilized nanofluids and their characteristics can increase the heat exchange rate of heat exchangers which are generally used in pipeline engineering. In order to help newcomers and researchers in this field recognize the potential research gap, this review study seeks to provide the latest research and development on stable nanofluids and their applications in pipeline engineering. Due to the lack of understanding of the mechanism of nanofluid at the atomic level, many experimental studies are needed to consider several important issues such as particle migration, agglomeration, and stability.

Author details

Sumit Kumar Singh
Department of Mechanical Engineering, Indian Institute of Technology (B.H.U.), Varanasi, India

*Address all correspondence to: sumitkrs.rs.mec15@itbhu.ac.in

References

[1] Xian HW, Sidik NAC, Saidur R. Impact of different surfactants and ultrasonication time on the stability and thermophysical properties of hybrid nanofluids. International Communications in Heat and Mass Transfer. 2020;**110**:104389

[2] Almanassra IW, Manasrah AD, Al-Mubaiyedh UA, Al-Ansari T, Malaibari ZO, Atieh MA. An experimental study on stability and thermal conductivity of water/CNTs nanofluids using different surfactants: A comparison study. Journal of Molecular Liquids. 2019;**19**:30397-30396

[3] Cacua K, Buitrago-Sierra R, Pabon E, Gallego A, Zapata C, Herrera B. Nanofluids stability effect on a thermosyphon thermal performance. International Journal of Thermal Sciences. 2020;**153**:106347

[4] Xu Q, Liu L, Feng J, Qiao L, Yu C, Shi W, et al. A comparative investigation on the effect of different nanofluids on the thermal performance of two-phase closed thermosyphon. International Journal of Heat and Mass Transfer. 2020;**149**:119189

[5] Said Z, Abdelkareem MA, Rezk H, Nassef AM, Atwany HZ. Stability, thermophysical and electrical properties of synthesized carbon nanofiber and reduced-graphene oxide-based nanofluids and their hybrid along with fuzzy modeling approach. Powder Technology. 2020;**364**:795-809

[6] Kazemi I, Sefid M, Afrand M. A novel comparative experimental study on rheological behavior of mono & hybrid nanofluids concerned graphene and silica nano-powders: Characterization, stability and viscosity measurements. Powder Technology. 2020;**366**:216-229

[7] Ouikhalfan M, Labihi A, Belaqziz M, Chehouani H, Benhamou B, Sarı A, et al. Stability and thermal conductivity enhancement of aqueous nanofluid based on surfactant-modified TiO2. Journal of Dispersion Science and Technology. 2020;**41**:374-382

[8] Siddiqui F, Tso C, Chan K, Fu S, Chao CY. On trade-off for dispersion stability and thermal transport of Cu-Al2O3 hybrid nanofluid for various mixing ratios. International Journal of Heat and Mass Transfer. 2019;**132**:1200-1216

[9] Akhgar A, Toghraie D. An experimental study on the stability and thermal conductivity of water-ethylene glycol/TiO2-MWCNTs hybrid nanofluid: Developing a new correlation. Powder Technology. 2018;**338**:806-818

[10] Choi TJ, Jang SP, Kedzierski M. Effect of surfactants on the stability and solar thermal absorption characteristics of water-based nanofluids with multi-walled carbon nanotubes. International Journal of Heat and Mass Transfer. 2018;**122**:483-490

[11] Das PK, Mallik AK, Ganguly R, Santra AK. Stability and thermophysical measurements of TiO2 (anatase) nanofluids with different surfactants. Journal of Molecular Liquids. 2018;**254**:98-107

[12] Gao Y, Wang H, Sasmito AP, Mujumdar AS. Measurement and modeling of thermal conductivity of graphene nanoplatelet water and ethylene glycol base nanofluids. International Journal of Heat and Mass Transfer. 2018;**123**:97-109

[13] Kuang W, Saraji S, Piri M. A systematic experimental investigation

on the synergistic effects of aqueous nanofluids on interfacial properties and their implications for enhanced oil recovery. Fuel. 2018;**220**:849-870

[14] Muthoka MJ, Xuelai Z, Xioafeng X. Experimental investigation on supercooling, thermal conductivity and stability of nanofluid based composite phase change material. Journal of Energy Storage. 2018;**17**:47-55

[15] Said Z, Allagui A, Abdelkareem MA, Alawadhi H, Elsaid K. Acid-functionalized carbon nanofibers for high stability, thermoelectrical and electrochemical properties of nanofluids. Journal of Colloid and Interface Science. 2018;**18**:30190-30195

[16] Alawi OA, Mallah AR, Kazi SN, Sidik NAC, Najafi G. Thermophysical properties and stability of carbon nanostructures and metallic oxides nanofluids. Journal of Thermal Analysis and Calorimetry. 2019;**135**:1545-1562

[17] Akbari A, Saidi MH. Experimental investigation of nanofluid stability on thermal performance and flow regimes in pulsating heat pipe. Journal of Thermal Analysis and Calorimetry. 2019;**135**:1835-1847

[18] Cacua K, Ordoñez F, Zapata C, Herrera B, Pabón E, Buitrago-Sierra R. Surfactant concentration and pH effects on the zeta potential values of alumina nanofluids to inspect stability. Colloids and Surfaces, A: Physicochemical and Engineering Aspects. 2019;**583**:123960

[19] Etedali S, Afrand M, Abdollahi A. Effect of different surfactants on the pool boiling heat transfer of SiO2/deionized water nanofluid on a copper surface. International Journal of Thermal Sciences. 2019;**145**:105977

[20] Giwa SO, Sharifpur M, Goodarzi M, Alsulami H, Meyer JP. Influence of base fluid, temperature, and concentration on the thermophysical properties of hybrid nanofluids of alumina–ferrofluid: Experimental data, modeling through enhanced ANN, ANFIS, and curve fitting. Journal of Thermal Analysis and Calorimetry. 2020;**6**:4149-4167

[21] Kazemi I, Sefid M, Afrand M. Improving the thermal conductivity of water by adding mono & hybrid nano-additives containing graphene and silica: A comparative experimental study. International Communications in Heat and Mass Transfer. 2020;**116**:104648

[22] Boroomandpour A, Toghraie D, Hashemian M. A comprehensive experimental investigation of thermal conductivity of a ternary hybrid nanofluid containing MWCNTs- titania-zinc oxide/water water ethylene (80:20) as well as binary and mono nanofluids. Synthetic Metals. 2020;**268**:116501

[23] Ding J, Zhao H, Yu H. Graphene nanofluids based on one-step exfoliation and edge-functionalization. Carbon. 2021;**171**:29-35

[24] Gul T, Firdous K. The experimental study to examine the stable dispersion of the graphene nanoparticles and to look at the GO–H2O nanofluid flow between two rotating disks. Applied Nanoscience. 2018;**8**:1711-1727

[25] Li X, Zou C, Chen W, Lei X. Experimental investigationof β-cyclodextrinmodified carbon nanotubesnanofluids for solar energy systems: Stability, optical properties and thermal conductivity. Solar Energy Materials & Solar Cells. 2016;**157**:572-579

[26] Rahimi K, Riahi S, Abbasi M. Effect of host fluid and hydrophilicity of multi-walled carbon nanotubes on stability and CO2 absorption of amine-based and water-based nanofluids. Journal of

Environmental Chemical Engineering. 2019;**19**:30703

[27] Vozniakovskii AA, Kol'tsova TS, Voznyakovskii AP, Kumskov AL, Kidalov SV. Powder hybrid nanomaterial: Detonation nanodiamonds – Carbon nanotubes and its stable reversible water nanofluids. Journal of Colloid and Interface Science. 2020;**565**:305-314

[28] Asadi A, Alarifi IM, Foong LK. An experimental study on characterization, stability and dynamic viscosity of CuO-TiO2/water hybrid nanofluid. Journal of Molecular Liquids. 2020;**307**:112987

[29] Chen Z, Shahsavar A, Al-Rashed AA, Afrand M. The impact of sonication and stirring durations on the thermal conductivity of alumina-liquid paraffin nanofluid: An experimental assessment. Powder Technology. 2020;**360**:1134-1142

[30] Asadi A, Alarifid IM, Ali V, Nguyen HM. An experimental investigation on the effects of ultrasonication time on stability and thermal conductivity of MWCNT-water nanofluid: Finding the optimum ultrasonication time. Ultrasonics Sonochemistry. 2019;**58**:104639

[31] Ranjbarzadeh R, Kazerouni AM, Bakhtiari R, Asadi A, Afrand M. An experimental study on stability and thermal conductivity of water/silica nanofluid: Eco-friendly production of nanoparticles. Journal of Cleaner Production. 2018;**18**:32938

[32] Aberoumand S, Jafarimoghaddam A. Tungsten (III) oxide (WO3) – Silver/transformer oil hybrid nanofluid: Preparation, stability, thermal conductivity and dielectric strength. Alexandria Engineering Journal. 2018;**57**:169-174

[33] Dalkılıç AS, Yalçın G, Küçükyıldırım BO, Öztuna S, Eker AA, Jumpholkul C, et al. Experimental study on the thermal conductivity of water-based CNT-SiO2 hybrid nanofluids. International Communications in Heat and Mass Transfer. 2018;**99**:18-25

[34] Aberoumand S, Ghamari S, Shabani B. Energy and exergy analysis of a photovoltaic thermal (PV/T) system using nanofluids: An experimental study. Solar Energy. 2018;**165**:167-177

[35] Kakavandi A, Akbari M. Experimental investigation of thermal conductivity of nanofluids containing of hybrid nanoparticles suspended in binary base fluids and propose a new correlation. International Journal of Heat and Mass Transfer. 2018;**124**:742-751

[36] Keyvani M, Afrand M, Toghraie D, Reiszadeh M. An experimental study on the thermal conductivity of cerium oxide/ethylene glycol nanofluid: Developing a new correlation. Journal of Molecular Liquids. 2018;**2018**:30794

[37] Liu X, Wang X, Huang J, Cheng G, He Y. Volumetric solar steam generation enhanced by reduced graphene oxide nanofluid. Applied Energy. 2018;**220**:302-312

[38] Ranjbarzadeh R, Akhgar A, Musivand S, Afrand M. Effects of graphene oxide'silicon oxide hybrid nanomaterials on rheological behavior of water at various time durations and temperatures: Synthesis, preparation and stability. Powder Technology. 2018;**2018**:30406

[39] Sharafeldin M, Gróf G. Experimental investigation of flat plate solar collector using CeO2-water nanofluid. Energy Conversion and Management. 2018;**155**:32-41

[40] Zeng J, Xuan Y. Enhanced solar thermal conversion and thermal conduction of MWCNT-SiO2/Ag

binary nanofluids. Applied Energy. 2018;**212**:809-819

[41] Gulzar O, Qayoum A, Gupta R. Experimental study on stability and rheological behavior of hybrid Al2O3-TiO2 Therminol-55 nanofluids for concentrating solar collectors. Powder Technology. 2019;**352**:436-444

[42] Alarifi IM, Alkouh AB, Ali V, Nguyen HM, Asadi A. On the rheological properties of MWCNT-TiO2/oil hybrid nanofluid: An experimental investigation on the effects of shear rate, temperature, and solid concentration of nanoparticles. Powder Technology. 2019;**355**:157-162

[43] Akram N, Sadri R, Kazi SN, Ahmed SM, Zubir MNM, Ridha M, et al. An experimental investigation on the performance of a flat-plate solar collector using eco-friendly treated graphene nanoplatelets–water nanofluids. Journal of Thermal Analysis and Calorimetry. 2019;**138**:609-621

[44] Sharafeldin M, Grof G. Efficiency of evacuated tube solar collector using WO3/Water nanofluid. Renewable Energy. 2019;**134**:453-460

[45] Chen W, Zou C, Li X. Application of large-scale prepared MWCNTs nanofluids in solar energy system as volumetric solar absorber. Solar Energy Materials and Solar Cells. 2019;**200**:109931

[46] Ali N, Teixeira JA, Addali A. Aluminum nanofluids stability: A comparison between the conventional two-step fabrication approach and the controlled sonication bath temperature method. Journal of Nanomaterials. 2019:3930572, https://doi.org/10.1155/2019/3930572

[47] Mahbubul I, Elcioglu E, Amalina M, Saidur R. Stability, thermophysical properties and performance assessment of alumina–water nanofluid with emphasis on ultrasonication and storage period. Powder Technology. 2019;**345**:668-675

[48] Mahyari AA, Karimipour A, Afrand M. Effects of dispersed added Graphene Oxide-Silicon Carbide nanoparticles to present a statistical formulation for the mixture thermal properties. Physica A. 2019;**521**:98-112

[49] Chen W, Zou C, Li X, Liang H. Application of recoverable carbon nanotube nanofluids in solar desalination system: An experimental investigation. Desalination. 2019;**451**:92-101

[50] Okonkwo EC, Wole-Osho I, Kavaz D, Abid M. Comparison of experimental and theoretical methods of obtaining the thermal properties of alumina/iron mono and hybrid nanofluids. Journal of Molecular Liquids. 2019;**292**:111377

[51] Teruel M, Aguilar T, Martínez-Merino P, Carrillo-Berdugo I, Gallardo-Bernal JJ, Gómez-Villarejo R, et al. 2D MoSe2-based nanofluids prepared by liquid phase exfoliation for heat transfer applications in concentrating solar power. Solar Energy Materials and Solar Cells. 2019;**200**:109972

[52] Li Z, Asadi S, Karimipour A, Abdollahi A, Tlili I. Experimental study of temperature and mass fraction effects on thermal conductivity and dynamic viscosity of SiO_2-oleic acid/liquid paraffin nanofluid. International Communications in Heat and Mass Transfer. 2020;**110**:104436

[53] Geng Y, Al-Rashed AAAA, Mahmoudi B, Alsagri AS, Shahsavar A, Alebizadehsardari P. Characterization of the nanoparticles, the stability analysis and the evaluation of a new hybrid nano-oil thermal conductivity. Journal of Thermal Analysis and Calorimetry. 2020;**139**:1553-1564

[54] Li Y, Kalbasi R, Nguyen Q, Afrand M. Effects of sonication duration and nanoparticles concentration on thermal conductivity of silica-ethylene glycol nanofluid under different temperatures: An experimental study. Powder Technology. 2020;**367**:464-473

Chapter 9

The Importance of Government Support for Pipeline Network Construction

Satoru Hashimoto

Abstract

This chapter describes that, focusing on economics theory, a country needs its government support to construct a pipeline network throughout the country. In Japan, before deregulation, vertical integrated companies (gas utilities) provided natural gas to customers in their monopoly area respectively. When the companies transport gas into their own areas, the companies choose to construct pipelines or to use LNG tank trucks from the sight of their strategies. Focusing on long-term uncertainty, short-term uncertainty, and locations, this chapter analyzed the factors of pipeline constructions. The results indicate that if there is another gas utility near a company, then the company construct a pipeline to the gas utility to transport gas. In contrast, if there are no neighbor utilities, the company tends to purchase gas via LNG tank truck. This means gas companies do not construct a pipeline network, or do not try to do it, but construct point to point pipelines. Therefore, without government supports, a pipeline network would not be constructed throughout the country.

Keywords: natural gas, transaction cost economics, Probit model, pipeline networks, LNG tank trucks

1. Introduction

This chapter considers the importance of government support for pipeline network construction focusing on Japan's natural gas industry from the perspective of economics, in particular, transaction cost economics and the organizational forms of gas local distribution utilities.

1.1 Overview of natural gas import

Before proceeding to outline this study, this section provides an overview of the domestic natural gas supply chain. Nearly all of Japan's natural gas requirements have been imported from overseas via LNG tankers. In 2020, 92.02% of all-natural gas was imported as LNG, 4.36% was produced from gas domestic fields, and the remainder was generated from imported petroleum-based gas (Agency for Natural Resources and Energy, Gas Market Division, 2021). Tokyo Gas and Tokyo Electric Power

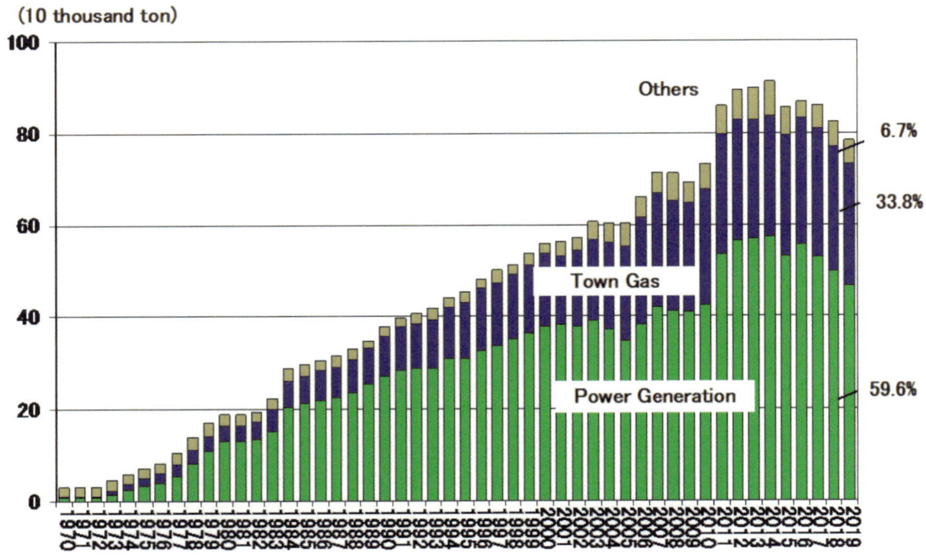

Figure 1.
Natural gas use in Japan (source: Ministry of Economics, trade, and industry).

Company (TEPCO) began to import liquefied natural gas (LNG) in 1969. Since then, increasing gas consumption has resulted in increasing imports of LNG. **Figure 1** shows domestical natural gas use, and indicates that approximately 60% of LNG is consumed for power generation, and "Town Gas" is basically used as cooking and heating by end users.

In recent years, many companies have started to import LNG, for example, the three major incumbents, Tokyo Gas, Osaka Gas, Toho Gas, and following incumbents with LNG terminals (Saibu Gas, Shizuoka Gas, and Hokkaido Gas). In addition, both upstream companies and power generation companies also import LNG. Here, the term 'upstream company' refers to a company that specializes in the production and transportation of energy, such as Japan Petroleum Exploration Company Limited (JAPEX) and INPEX Corporation, while power generation companies include Tokyo Electric Power Corporation (TEPCO), Kansai Electric Power Corporation (KEPCO), and Chubu Electric Power Corporation (CEPCO).

1.2 Retail markets and pipeline networks

The gas retail market has two main features. First, the companies are classified into two categories based on ownership structure. As of March 2021, 173 utilities were private companies while 20 were municipality-owned companies. Of the 173 privately owned utilities, 12 were listed companies. Second, the size of these firms varies significantly. As can be seen in **Table 1**, the maximum revenue is USD 12,586,010,000 (Tokyo Gas), while the minimum is USD 294,740. The large and medium-sized incumbents are involved in production (import), transmission, and distribution, that is, they are vertically integrated utilities. Meanwhile, there are medium- and small-sized incumbents that only sell gas to consumers. These incumbents purchase natural gas from upstream companies via pipelines or LNG tank trucks.

One of the reasons for this situation is that a regional monopoly policy has previously been enforced in the natural gas industry. As far as we can see the natural gas

	Revenue	Production	Customers	Tangible Assets
(Unit)	USD	1000 MJ	People	USD
Average	171,474,800	8,421,867	188,521	135,625,510
S.D.	1,052,895,090	55,953,815	1,037,167	731,440,770
Min.	294,740	5590	510	239,650
Max	12,586,010,000	691,882,206	12,208,885	8,989,040,000

(S.D.: Standard Deviation, Source: Gas Business Annual Report, 2015)
(1USD = 100JPY)

Table 1.
Basic information relating to gas distribution utilities (2015).

industry after 1945, gas utilities had never been integrated politically, whereas electric utilities were integrated into 10 groups. Also, the spread of LPG use (liquified petroleum gas) since 1945 had greatly influenced the establishment of small sized natural gas utilities. Therefore, the natural gas industry has many utilities (incumbents) and large difference of the largest and smallest utilities. This policy has also affected the characteristics of the natural gas distribution network. **Figure 2**, in which the trunk pipeline networks[1] are depicted, shows that the trunk pipeline networks are quite poor, that is, the coverage area of networks is narrow and they are not enough connected, although the natural gas consumption in 2016 was 111.2 billion m^3 which represented about 3.1% of global consumption, and Japan is the largest LNG importing country in the world[2].

As is well known, LNG is transformed into natural gas by regasification facilities at or close to LNG terminals. Each large incumbent typically constructed LNG terminals at a sea port close to cities with large populations in its monopolistic supply area and constructed new pipelines after estimating the profits that would be generated by the additional investment. Meanwhile, middle or small incumbents purchase LNG via tank truck or natural gas via pipelines from upstream companies. In the former case, incumbents constructed gasified facilities, and in the latter case, incumbents constructed trunk pipelines to neighbor suppliers. According to the Gas Business Annual Report (2015), total transportation volumes via pipelines and tank trucks were approximately 1735 billion MJ and 1324 billion MJ, respectively, in 2015. Even Tokyo Gas, the largest gas company, uses LNG tank trucks to haul natural gas in a part of its supply area. Consequently, pipeline networks radiated outwards from the 35 existing LNG terminals, and becomes narrow networks. Hence, the pipeline networks terminate in each region, and there are insufficient trunk pipelines connecting the various regions. Incumbents with a vertically integrated structure would not have sufficient incentive to connect their trunk pipelines with those of other incumbents. In fact, there was no pipeline connecting Tokyo and Osaka, a total distance of approximately 2450 km, until December 2015 (Source: Gas Business Annual Report, 2015).

[1] The Japan Gas Association and regulatory authority do not classify pipelines into transmission and distribution pipelines. Instead, they classifies them into high-, medium-, and low-pressure pipelines. Trunk pipelines in **Figure 2** mean high-pressure pipelines.

[2] See *BP* Statistical *Review of World Energy, June 2017.*

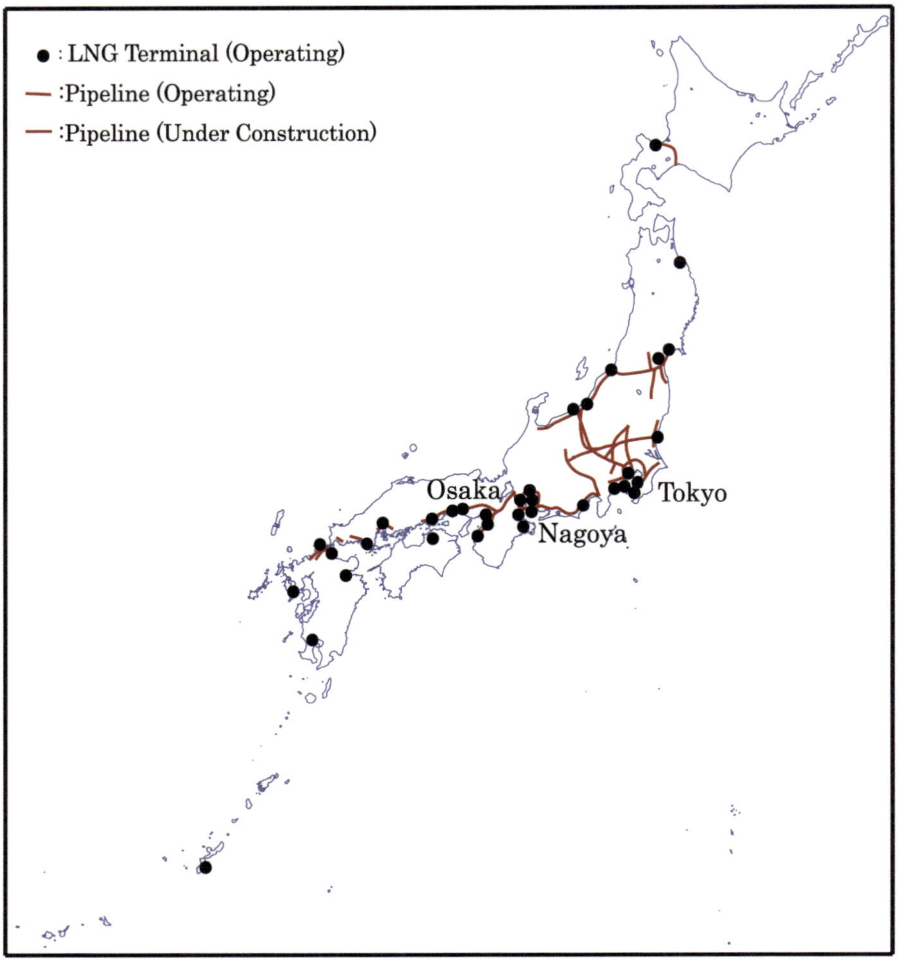

Figure 2.
Pipeline networks and LNG terminals in 2016 (source: The Japan gas association).

Figure 3 shows the LNG terminals and pipeline networks of Osaka Gas, with the supply area shaded red. It is easy to see that high-pressure (red) and medium-pressure (green) pipelines spread outward from the two LNG terminals. **Figure 3** also shows that the Japanese pipeline network is sparse. The supply area for Gojo Gas is shaded yellow. Although Gojo city in the Nara prefecture is located approximately 41.8 km southeast of Osaka city, there are no pipelines connecting the Osaka Gas and Gojo Gas.

Sadorsky (2001) [1] indicates that it is difficult to introduce product differentiation in relation to natural gas, therefore, gas suppliers (utilities) are most likely to face price competition. Weir (1999) [2] describes that if an incumbent owns both transport and distribution facilities, it would be a barrier to entry for entrants that only have third-party access to pipelines. Hence, the UK government introduced unbundling regulations to increase the number of new entrants and improve competition in the UK retail market.

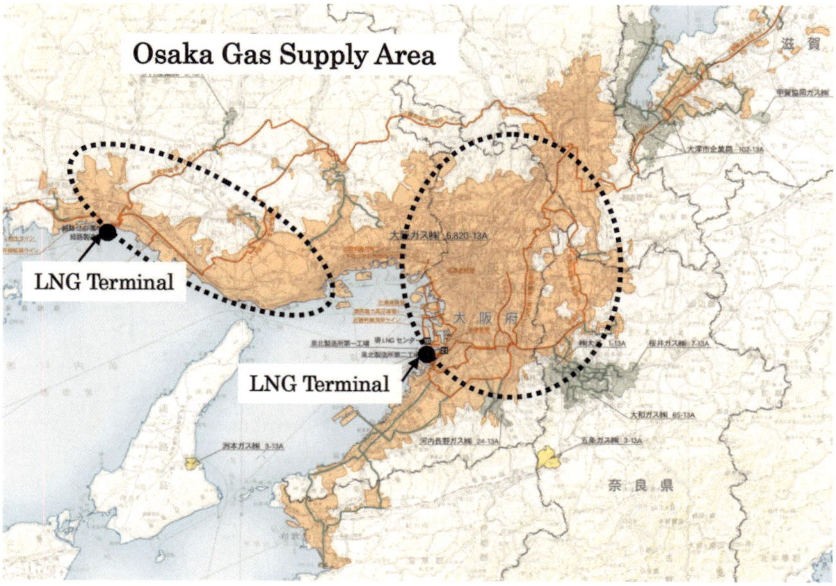

Figure 3.
Osaka gas pipeline network and LNG terminals (source: Ministry of Economics, trade, and industry).

1.3 Natural gas supply chains

The natural gas supply chain varies according to their historical and geographical characteristics.

Sailer et al. (2009) [3] define a natural gas supply chain as following six stages: exploration, extraction, production, transportation, storage, and distribution. In this study, the supply chain is simplified three stages: import, gasification, and distribution (**Figure 4**). The "import" activities include transportation from overseas, exploration, extraction, and production because it focuses on the process of delivering imported gas to end users. "gasification" activities involve procedures to gasify LNG into natural gas and to transport gas from upstream companies to distribution utilities. "distribution" activities include both storage and distribution into end users. In general, Upstream companies operate "import" activities while downstream companies operate "distribution" activities. Regarding "gasification" activities, in some cases, the upstreams operate, and in the other cases the downstreams do. Basically, no transportation companies with pipelines or regasification facilities that are independent of

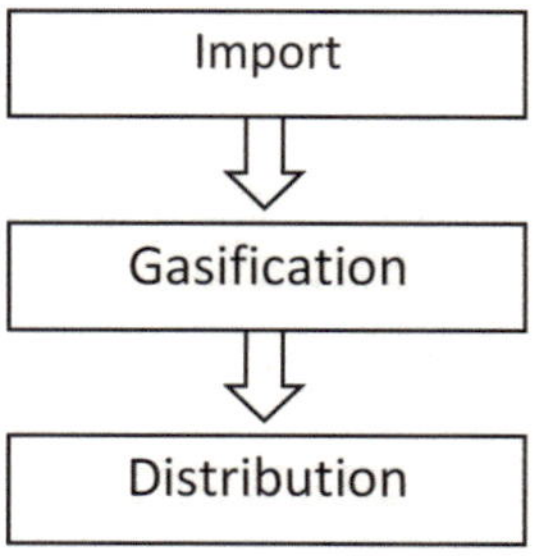

Figure 4.
The natural gas supply chain.

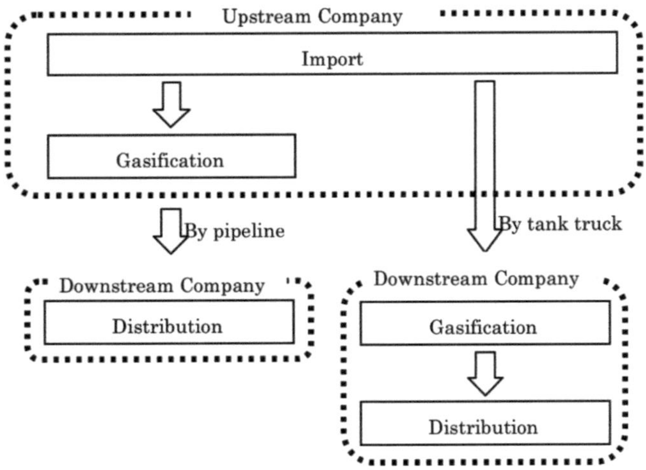

Figure 5.
Organizational structures.

the upstreams and downstreams exist in Japan (**Figure 5**). Thus, either an upstream or a downstream company needs to shoulder the responsibility for regasification activities to re-gasify LNG and transportation.

Unlike in the United States and EU countries, the Japanese Government (regulatory authority) had never enforced unbundling regulations that prohibit management of both transportation (including import) and distribution activities until 2020, but since April of 2022, the government has introduced the regulations into three largest companies. Besides, almost all upstream companies basically own gas storage, while local distribution utilities (downstream companies) have to undertake the responsibility for stable gas supply to end users. Also, the Government had authorized local distribution utilities to provide natural gas on the principle of a natural monopoly, permitting the utility a business license in the permitted area. These utilities are obligated to provide natural gas to end users in their own area safely and continually, however, the government had not imposed any relevant regulations on gasification activities.

Many upstream companies have both gasification facilities and trunk pipelines to provide into local distribution utilities with natural gas through pipelines or with LNG using tank trucks (**Figure 5**). Because raw commodities such as natural gas is impossible with product differentiation, the best performing natural resource companies are generally the lowest cost producers (Sadorsky 2001) [1]. Therefore, taking into account cost minimization and management strategies, the utilities purchase natural gas directly by joining a pipeline from its own facility to trunk pipelines owned by upstream companies[3], or purchase LNG by tank trucks. If distribution utilities purchase LNG directly, then they need to construct in-house gasification facilities to provide natural gas into end users. As a result, some downstream utilities have in-house gasification facilities, while the others do not (**Figure 5**).

---

[3] Many distribution utilities can purchase natural gas via pipelines from not only upstream companies but neighboring distribution utilities (downstream companies).

1.4 Government policy and utility's decision

A utility's decision on whether or not it needs to establish gasification facilities is critical to that utility's attempts to manage its economic performance. Besides, three historical or crucial circumstances might affect this decision: the Integrated Gas Family 21 plan (IGF 21) issued by the Ministry of International Trade and Industry in 1990[4], official network plans by the government, and managerial uncertainties.

Next, regarding pipeline network plans, the government has not yet made official plans for a pipeline network, nor has it provided financial aid for its construction, although both retail prices and supply areas have been regulated for a long time. When incumbents implement construction of their own pipelines, they need to raise the money for its construction, and, the decision to implement it depends upon long-term demand and managerial efficiencies. In the case where a massive amount of gas is transported, pipelines are superior to LNG tank truck, but, the former option requires huge capital expenditure to build pipeline facilities. When an incumbent encounters large uncertainties related to the weather conditions (meteorological conditions) or a volatile industrial demand that is affected by the economic conditions, it tends to refrain from an investment in a trunk pipeline even if large demand is expected. In these cases, the incumbent selects to purchase LNG via tank trucks. Thus, the government has never been strongly concerned with pipeline construction.

Also, managerial uncertainties might affect the vertical integration choice. Stable procurement is an indispensable part of distribution utilities. However, it might be difficult to implement an obligation to sustain security of supply for a long time. This is because even if managerial uncertainties are large, the distribution utilities have to continue providing natural gas in constant and sufficient quantities for a long time. To decrease the uncertainties, some utilities strive to purchase gas from plural wholesalers to maintain multiple supply chains, while others set up multiple natural gas storage tanks. Hence, pipeline construction would be affected by political issues, uncertainties, stable procurement. As a manner to explore pipeline construction factors, this study, focusing on a transaction cost economics theory, estimates the transaction cost empirically, and then considers the importance of comprehensive construction policies.

2. Theoretical background

2.1 Transaction cost economics

Here, the context of Transaction Cost Economics (TCE) is defined.

Coase (1937) [4] predicted existence in external costs between two firms, and for a single firm, internal costs exist between its divisions. The concept of external and internal costs was considered to be one of the significant factors when an entrepreneur determines a firm's boundaries. When engaging in business transactions, a firm has a strong incentive to integrate with another firm that has significantly high external costs. In contrast, if the external costs between the two firms are not very high, then the former firm does not have a strong incentive to integrate with the latter firm, though would continue to do business with it.

---

[4] See The Japan Gas Association, https://www.gas.or.jp.

The external and internal cost concept defined by Coase (1937) [4] was developed into transaction cost economics by Williamson (1975, 1985, 1995) [5–7], who defined the term "invisible costs" as "transaction costs", and explained the origin of transaction costs based on three factors: (a) uncertainty, (b) relationship-specific assets, and (c) frequency (Williamson, 1985) [6].

There are several notable papers based on transaction cost views of vertical integration in the manufacturing industry. First, Monteverde and Teece (1982) [8] analyze asset specificity of GM and Ford, and found that the probability of vertical integration between a parent company and a subsidiary might rise because the manufacture of parts which needs advanced technologies in automobiles tends to become relationship specific assets. Second, Masten et al. (1989) [9], separating specific assets into human assets and physical assets, insist that human assets affect vertical integration more than physical assets. Third, Walker and Weber (1984,1987) [10, 11] show that, focusing on the uncertainty, when the uncertainty to get manufacturing parts becomes higher the probability for vertical integration also becomes higher.

Regarding empirical analyses, Levy (1985) [12] estimates the boundaries of firm by using 67firms' data (37industries), and puts asset specificity as R&D investment, and moreover puts uncertainty as variance of sales. As a seminal work for transaction cost economics of power generation industry, Joskow (1985, 1988) [13, 14] found that power generation plants incline to be constructed close to mining pits, and that vertical integration between plants and pits, and long-term contracts, were widely practiced. Crocker and Masten (1996) [15] investigated the organizational forms of public utilities in the United States.

Shelanski and Klein (1995) [16] reviewed many empirical literatures related to transaction cost economics theory, and then concluded that they have conspicuously consistent with predictions from the theory. In contrast, David and Han (2004) [17] and Carter and Hodgson (2006) [18], from traditional literature surveys, found that asset specificity and uncertainty have received considerable scrutiny or commonly examined, whereas frequency has not. Hence, they concluded that some literatures have produced results that the transaction cost framework would not predict.

Sheravani et al. (2007) [19] noted the importance of the relationship between transaction costs and market power. They suggested that high market power appears to provide safeguards to a firm using nonintegrated channels not envisioned in predictions from transaction cost economics. Furthermore, they argued that firms with high market power are likely to have significant monitoring and surveillance capabilities, can exercise legitimate authority, and offer diverse incentives to associated channel members[5].

2.2 Organizations and LNG supply chains

This section introduces several related literatures on LNG supply chains. Xunpeng, (2016) [20] points out that almost all the incumbent gas companies in Asian have vertically integrated supply chains. Lee et al. (1999) [21] found that KOGAS (the

[5] To evaluate market power, this study measured the Hyfindal Hussuman Index (HHI) based on sales volume. In 2010, the HHI value of gas distribution utilities was 2037, which indicated that the market power of the industry was not very high. However, Tokyo Gas, the largest gas utility, had 30% of the market share, and Osaka Gas, the second largest utility, had a 20% market share. In addition, these utilities each had three types of operations activity, from import to distribution. Therefore, this study estimated the transaction costs excluding these two largest utilities.

Korean national firm) have a lower productivity level compared to firms with acquiring their gas through pipelines because it depends on LNG import, which requires additional capital facilities for shipping, storage, and regasification. As been described by Vivoda (2014a) [22], a number of international LNG trade are dominated by long-term contracts. This is because the trading companies need the huge capital costs including liquefaction and regasification facilities and the inherent inflexibility in the value chain required contractual arrangements to protect both the suppliers and the buyers. In contrast, Cabalu (2010) [23] and Hartley (2013) [24] described that technological innovations make LNG transportation costs decrease significantly and LNG import and export volume were gradually increasing. Also, Gkonis and Psaraftis (2009) [25] indicate that LNG shipping markets are basically oligopolistic, and then suggested that competing companies have to consider a transportation capacity in the LNG shipping market. Vivoda (2014b) [26] points out that it is important for Japan and Korea to elaborate diverse LNG strategies to import LNG smoothly.

Turner and Johnson (2017) [27] denote that LNG trade is superior to pipeline transportation, and then point out that importers and exporters can easily send and receive gas to any locations with liquefaction and regasification facilities when LNG trade is possible. Xunpeng (2016) [20] and Hashimoto (2020) [28] describe the prospect in Asian LNG spot and hub markets.

3. Methodology and data

Here, this section describes the application of transaction cost economics theory into gas utilities. According to this theory, transaction costs consist of external and invisible costs between two firms, while internal costs are composed from invisible costs between two divisions in a single firm (**Figure 6**). A firm basically assumes to determine the choice whether or not it selects vertical integration, based on the transaction cost economics theory. **Figure 6** illustrates the application of transaction cost economics. When transaction costs exceed internal costs, then firm B would merge with firm A or acquire it. This study defines this type of consolidation as a vertical integration. In contrast, when internal costs exceed transaction costs, firm B would not merge with firm A.

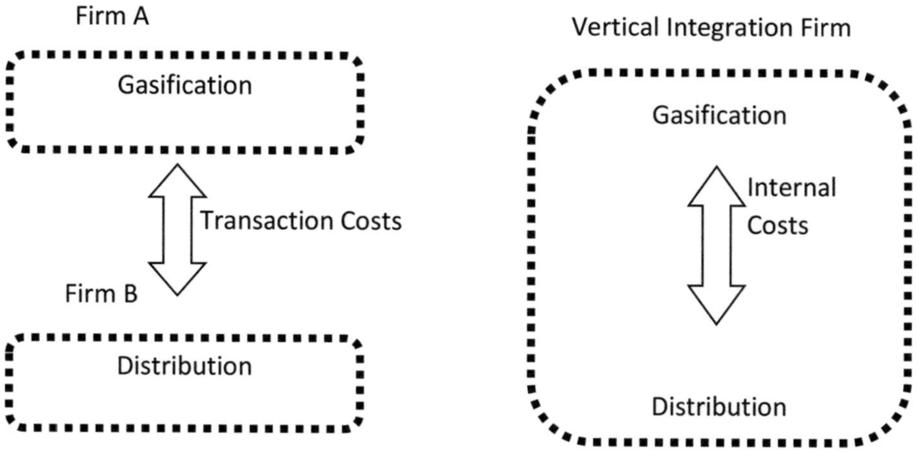

Figure 6.
Schematic of application of transaction costs and internal costs.

Regarding empirical estimations, this theory has two prudent treatments. First, Williamson (1985, p.20) [6] indicates the existence of ex ante and ex post transaction costs, and Monteverde and Teece (1982) [8] investigated ex ante transaction costs. This study also attempts to investigate ex ante transaction costs. Second, although both transaction costs and internal costs should be observed directly, it would be impossible to measure internal costs precisely. However, because transaction cost economics theory expects a positive correlation between transaction costs and incentive for vertical integration this study estimates only transaction costs to analyze the incentive.

Williamson (1985) [6] explains uncertainty, relationship-specific assets, and frequency as transaction costs. Many empirical analyses treat with uncertainty and relationship-specific assets [8–12]. Frequency is not analyzed because of the difficulty in estimation. This study also adopts the two components; uncertainty and relationship-specific assets. Regarding uncertainty, this study divides it into long-term and short-term components for its importance to the natural gas industry. In distribution utilities, because growth rate and demand fluctuation would be the main components of uncertainty, this study defined sales volume, number of customers, and average revenue growth rate as sources of long-term uncertainty. Also, Gas demand and underpinning sales would be affected by seasonal factors, which are defined as monthly sales variance and the inventory rate. These are components of short-term uncertainty in this analysis[6].

Meanwhile, relationship-specific assets are classified into the assets by site specificity, physical asset specificity, human asset specificity, and dedicated assets by Williamson (1985) [6]. This study employs physical and site specificities, and does not analyze physical assets, human assets, and dedicated assets. For physical asset specificity, gas utilities comprise public and private administrations. Public gas utilities are expected to receive aid from municipalities when they face bankruptcy. Therefore, they may decrease transaction costs for related to relationship-specific assets.

To scrutinize long-term uncertainty, short-term uncertainty, and site specificity, the Integration equation is assumed by means of the methodology of Levy (1985) [12] and Wang and Mogi (2017) [29];

$$Integration = f(LU, SU, SS) \tag{1}$$

where LU, SU, and SS are long-term uncertainty, short-term uncertainty, and site specificity, respectively. If transaction costs increase, the value of Integration would become high. Hence, the high value means high incentive to integrate.

The dependent variable, "vertical integration (VI)", represents whether or not a utility has gasification facilities. When a distribution utility has gasification facilities, it means that transaction costs exceed internal costs. Hence, "vertical integration (VI)" is adopted as the dependent variable. The dependent variable (x) represents whether or not a utility has gasification facilities.

Table 2 shows the definitions of dependent and independent variables, including the expected sign. RGR, SAL, CUS, ASS, and PRO are defined as components of long-term uncertainty, and SDR, SDM, SVV, AVI, and HHR are defined as components of short-term uncertainty. SSD and PUD are adopted as components of site specificity. The data sources are gas business annual reports, and SSD was obtained from the natural gas supply area map (agency for natural resources and energy).

In this study, the integration eq. (1) was defined as:

[6] More detailed information is described by Hashimoto (2021) [30].

Variable	Definition	Expected sign
VI	If the utility possesses regasification facilities, then assign 1. Otherwise assign 0.	
RGR	Average revenue growth rate from 2006 to 2015 (absolute value)	+
PRO(LU)	Production volume in 2015 (1000 MJ)	+
ASS(LU)	Utility's tangible assets in 2015 (1000 JPY)	+
SAL(LU)	Utility's sales volume in 2015 (1000 MJ)	+
CUS(LU)	Number of customers in utility's monopoly area (People)	+
SVV(SU)	Standard deviation of utility's monthly sales volume from January to December divided by the whole sales volume in 2015 $\left(\frac{\text{Monthly sales volume variance from January to December in 2015}}{\text{The sales volume in 2015}}\right)$	+
SDM(SU)	Standard deviation of utility's monthly sales volume from January to December	+
AVI(SU)	Average natural gas inventory for the past 3 years (absolute value) $\left(\frac{\text{Product}-\text{Sales}}{\text{Product}}\right)$	+
HHR(SU)	Household rate $\left(\frac{\text{Household sales volume}}{\text{Wholesales volume}}\right)$	—
SDR(SU)	Standard deviation of revenue for the 10-year period (2006–2015)	—
SSD(SS)	Site specificity dummy (2015) If a utility borders other distribution utilities or there are domestic natural gas fields in its own area, assign 1. Otherwise assign 0.	—
PUD(SS)	Public utility dummy (Public utility: 1, Private utility: 0)	—

Table 2.
Definition of variables (source; Hashimoto, 2021 [30]).

$$
\begin{aligned}
e^x &= \alpha\prod_l LU_l^{b_l}\cdot\prod_m SU_m^{c_m}\cdot\prod_n SS_n^{d_n}\\
\ln e^x &= \ln\alpha\prod_l LU_l^{b_l}\cdot\prod_m SU_m^{c_m}\cdot\prod_n SS_n^{d_n}\\
x &= \beta+\sum_l b_l\ln LU_l+\sum_m c_m\ln SU_m+\sum_n d_n\ln SS_n,
\end{aligned}
\tag{2}
$$

where x means the probability of integration. LUl, SUm, and SSn respectively mean the l-th long-term uncertainty, m-th short-term uncertainty, and n-th site specificity, while α and β are constants. The high value of x means high incentive to integrate, while the low value means low incentive. **Table 3** shows the descriptive statistics.

4. Results

Table 4 shows the Probit model results. The results of model 1–6 indicate robustness. Because strong correlations among independent variables of long-term uncertainty (LU) might be shown, these variables were not applied simultaneously[7].

The three types of transaction costs will be discussed[8]. First, for long-term uncertainty, the coefficients of PRO, ASS, SAL, and CUS are not significant at 10% level.

[7] In general, strong correlations among variables are neither necessary nor sufficient to cause multicollinearity. More detailed information is described by Hashimoto (2021) [30].

[8] More detailed information and discussion is described in Hashimoto (2021) [30].

	Integration	RGR	PRO	ASS	SAL	CUS	SVV	SDM
average	0.624	0.043	8,462,922	13,534,961	7,445,264	189,432	0.018	87,643
S.D.	0.034	0.007	3,917,333	5,121,035	3,355,212	72,610	0.001	37,816
Min.	0	0.000	6451	23,965	6436	510	0.003	148
Max	1	1.015	691,882,206	898,904,000	577,580,996	12,208,885	0.043	6,516,900

(S.D.: Standard deviation)

Table 3.
Descriptive statistics (source; Hashimoto, 2021 [30]).

	Model 1	Model 2	Model 3	Model 4	Model 5	Model 6
Constant	2.656* (1.54)	1.923 (1.56)	2.622* (1.54)	2.396 (1.46)	2.369 (1.52)	2.622* (1.54)
RGR(LU)	−0.190* (0.10)	−0.189* (0.10)	−0.190* (0.10)	−0.186* (0.10)	−0.202* (0.11)	−0.190* (0.10)
PRO(LU)	0.142 (0.06)					
ASS(LU)		0.086 (0.07)				
SAL(LU)			0.017 (0.06)			
CUS(LU)				0.071 (0.06)		
SVV(SU)	0.474* (0.26)	0.555** (0.26)	0.478* (0.26)	0.533** (0.26)	0.517** (0.26)	
AVI(SU)	0.197*** (0.07)	0.192*** (0.69)	0.197*** (0.70)	0.186*** (0.07)	0.191*** (0.07)	0.197*** (0.07)
HHR(SU)	−0.132 (0.70)	−0.120 (1.82)	−0.129 (0.19)	−0.140 (0.18)	−0.103 (0.19)	−0.129 (0.19)
SDR(SU)					0.049 (0.06)	
SDM(SU)						0.017 (0.06)
SSD(SS)	−0.930*** (0.26)	−1.016*** (0.22)	−0.936*** (0.23)	−0.978*** (0.21)	−0.979*** (0.22)	−0.936*** (0.25)
PUD(SS)	−0.487* (0.30)	−0.514** (0.30)	−0.487* (0.30)	−0.463 (0.30)	−0.481 (0.30)	−0.487** (0.29)
R-squared	0.208	0.214	0.208	0.212	0.210	0.207
Log likelihood	−119.409	−118.551	−119.395	−118.868	−119.140	−119.395
Observations	205	205	205	205	205	205
Franction of collect predictions	0.68	0.70	0.68	0.69	0.69	0.68

*(Note: Standard errors are in parentheses, and ***, **, and * are significant at 1, 5, and 10%, respectively)*

Table 4.
Probit model results (source; Hashimoto, 2021 [30]).

RGR is significant at 10% in all models, but the sign was negative, not being consistent with the expected sign. Hence, the component of long-term uncertainty is required careful interpretation[9].

Second, for short-term uncertainty, AVI was significant at 1% in all models, and standard deviation of SVV was also significant at 5% or 10%. HHR, SDR and SDM were not significant at the 10% level. Regarding coefficients for AVI, the consideration for a causal relation is required carefully because it would be possible to interpret that utilities with possessing regasification facilities increase their inventory volume. This study concluded that short-term uncertainty cannot be strongly supported.

Third, the coefficients of SSD were significant at the 1% level in all models, and also, those of PUD were significant at 5% or 10% level in some models. Therefore, the existence of assets and site specificities are strongly recognized.

5. Conclusions

This chapter, to confirm the importance of comprehensive pipeline network plans by the government, examined whether or not local distribution utilities integrated gasification activities, in terms of transaction cost economics theory, and then found that the utilities prefer to purchase natural gas via pipelines when there are wholesalers or neighboring utilities that provide gas to end users, or there are natural gas fields in the vicinity. As a result of the utility's behavior, a broad pipeline network would not be built throughout the country.

Pipeline construction depending on distribution utilities affects the growth of broad pipeline networks. Because a utility prefers to construct a point-to-point pipeline between itself and a neighbor wholesaler the utility by no means construct a pipeline network by itself. Even if a number of utilities invest the point-to-point pipelines, a nationwide pipeline network cannot be built. Consequently, as seen in **Figure 2**, a broad pipeline network infrastructure has not been constructed throughout Japan. Therefore, this study concluded that we need the government support for pipeline network construction.

Table 5 shows the cases of pipeline investment. **Table 5** indicates that companies invest comparatively small distance pipelines that are under 100 km, and moreover, joint ownership is adopted by the two related firms such as distribution utilities and wholesalers[10]. The former would reinforce this study that distribution utilities tend to purchase gas from those companies when there are wholesalers or neighbor utilities in the vicinity. Meanwhile, the latter might bear out the necessity of government supports for nationwide pipeline networks. The government basically has not supported the companies by means of pipeline network planning, comprehensive infrastructure policies, or financial support. The reason why joint-ownership is adopted might be without any governmental supports. In other words, this might be indicating that the

---

[9] In fact, the relationship between firm size or sales volume and the risk of purchasing natural gas is unclear. More detailed discussion is described by Hashimoto (2021) [30].

[10] In **Table 5**, Ogijima-Kawasaki line [1], Minamifuji Line [3], Setouchi line [4], Koriyama line [5], Shizuhama line [6], Isewan Odan pipeline [8], and Mie-shiga line [9] are adopted a form of joint-ownership. Meanwhile, Chiba-Kashima line [7], Saito line [12], Ibaragi-Tochigi line [13], Furukawa-Maoka line [14], and Ibaragi line [15] are the projects of Tokyo Gas, and Himeji-Okayama line [10] and Amagasaki-Kugayama line [16] are the projects of Osaka Gas. Those projects have been invested by a single company because those pipelines are all in its own monopolistic area.

No.	Pipeline's name	Organizational form (Equity rate)	Year	Distance (Bore)
1	Ogijima-Kawasaki	Kawasaki Gas Pipeline Co. Ltd. {Tokyo Gas: 50%, Nihon Oil: 50%}	2004	5 km (400 ~ 500 mm)
2	Shizuoka Line Showa-Gotenba	INPEX. Co. Ltd.	2006	81 km (400 mm)
3	Minamifuji Line Fuji-Gotenba	Minamifuji Pipeline Co. Ltd. {INPEX, Shizuoka Gas, Tokyo Gas: 33% respectively}	2006	31 km (500 mm)
4	Setouchi Line Mizushima-Fukuyama	Setouchi Pipeline Co. Ltd. {Hiroshima Gas:80%,Fukuyama Gas:20%}	2006	40 km (300 mm)
5	Koriyama Line Shiroishi-Koriyama	INPEX co.:80%,Tohoku Electric Power:20% (Joint Ownership)	2007	95 km (400 mm)
6	Shizuhama Line Shizuoka-Hamamatsu	Shizuhama Pipeline Co. Ltd.(part of 76 km) {Shizuoka Gas: 50%, Chubu Gas: 50%}	2012	113 km (400 ~ 500 mm)
7	Chiba-Kashima Line	Tokyo Gas	2012	79 km (600 mm)
8	Isewan Odan Pipeline Chita,Kawagoe-Yokkaichi	Chubu Electric Power:50% Toho Gas: 50% (Joint Ownership) Undersea tunnel	2013	19 km (600 ~ 700 mm)
9	Mie-Shiga Line Hikone-Yokkaichi	Osaka Gas(20 km), Chubu Electric Power(40 km)	2014	60 km (600 mm)
10	Himeji-Okayama-Line	Osaka Gas Co. Ltd.	2014	86 m (600 mm)
11	Toyama Line Itoigawa-Toyama	INPEX. Co. Ltd.	2014	102 km (500 mm)
12	Saito Line	Tokyo Gas	2015	40 km (600 mm)
13	Ibaragi-Tochigi Line	Tokyo Gas	2015	81 km (600 mm)
14	Furukawa-Maoka Line	Tokyo Gas	2018	50 km (600 mm)
15	Ibaragi Line	Tokyo Gas	2021	92 km (600 mm)
16	Amagasaki-Kugayama Line	Osaka Gas	2028	49 km (600 mm)

Table 5.
The cases of pipeline investment.

government need to have some supports to build a nationwide pipeline network throughout the country.

In addition, the government decided to unbundle natural gas companies. The three largest incumbents (Tokyo gas, Osaka gas, and Toho Gas) were unbundled in April, 2022. While the unbundling regulation is expected to promote competition, it might also discourage pipeline investment. Hence, incentives for investment after the introduction of unbundling regulation might need to be considered.

Finally, this study explored the determinants of vertical integration and the importance of pipeline network plans. However, this study has not considered or investigated the characteristics of network externalities, a natural monopoly, economies of scale, and management strategies that cannot be evaluated by transaction cost economics[11]. They will be the focus on future work.

[11] See Baumol and Oates (1975) [31] and Sharkey (1982) [32].

Author details

Satoru Hashimoto
Faculty of Economics, Teikyo University, Tokyo, Japan

*Address all correspondence to: s-hashi@main.teikyo-u.ac.jp

References

[1] Sadorsky P. Risk factors in stock returns of Canadian oil and gas companies. Energy Economics. 2001;**23**: 17-28. DOI: 10.1016/S0140-9883(00)00072-4

[2] Weir C. Regulation and the development of competition in the U.K. gas supply industry. Review of Industrial Organization. 1999;**15**:135-147

[3] Sailer S, Selk E, Baumgart G. Vertical Integration in the Natural Gas Value Chain: Historical and Current Developments. Seminar Paper, GRIN Verlag GmbH; 2009

[4] Coase RH. The nature of the firm. Economica. 1937;**4**(16):386-405. DOI: 10.1111/j.1468-0335.1937.tb00002.x

[5] Williamson OE. Market and Hierarchies, Analysis and Antitrust Implications. The Free Press; 1975

[6] Williamson OE. The Economic Institutions of Capitalism. The Free Press; 1985

[7] Williamson OE. The Mechanisms of Governance. Oxford University Press; 1996

[8] Monteverde K, Teece DJ. Switching cost and vertical integration in the automobile industry. The Bell Journal of Economics. 1982;**13**(1):206-213. DOI: 10.2307/3003441

[9] Masten SE, Meehan JW Jr, Snyder EA. Vertical integration in the U. S. auto industry. Journal of Economic Behavior and Organization. 1989;**12**: 265-273

[10] Walker G, Weber D. A transaction cost approach to make-or-buy decisions. Administrative Science Quarterly. 1984; **29**(3):373-391. DOI: 10.2307/2393030

[11] Walker G, Weber D. Supplier competition, uncertainty, and make-or-buy decisions. The Academy of Management Journal. 1987;**30**(3): 589-596. DOI: 10.5465/256017

[12] Levy DT. The transactions cost approach to vertical integration: An empirical examination. The Review of Economics and Statistics. 1985;**67**(3): 438-445. DOI: 10.2307/1925972

[13] Joskow PL. Vertical integration and long-term contracts: The case of coal-burning electric generating plants. Journal of Law, Economics, & Organization. 1985;**1**(1):33-80

[14] Joskow PL. Asset specificity and the structure of vertical relationships: Empirical evidence. The Journal of Law Economics, & Organization. 1988;**4**(1): 95-117

[15] Crocker KJ, Masten SE. Regulation and administered contracts revisited: Lessons from transaction cost economics for public utility regulation. Journal of Regulatory Economics. 1996;**9**:5-39

[16] Shelanski HA, Klein PG. Empirical research in transaction cost economics: A review and assessment. Journal of Law, Economics & Organization. 1995;**11**(2): 335-361

[17] David RJ, Han SK. A systematic assessment of the empirical support for transaction cost economics. Strategic Management Journal. 2004;**25**:39-58. DOI: 10.1002/smj.359

[18] Carter R, Hodgson GM. The impact of empirical tests of transaction cost

economics on the debate on the nature of the firm. Strategic Management Journal. 2006;**27**:461-476. DOI: 10.1002/smj.531

[19] Sheravani TA, Frazier G, Challagalla C. The moderating influence of firm market power on the transaction cost economics model: An empirical test in a Forward Channel integration context. Strategic Management Journal. 2007;**28**:635-652

[20] Xunpeng S. Gas and LNG pricing and trading hub in East Asia: An introduction. Natural Gas Industry B. 2016;**3**:352-256. DOI: 10.1016/j.ngib.2016.12.008

[21] Lee JD, Park SB, Kim TY. Profit, productivity, and Price differential: An international performance comparison of the natural gas transportation industry. Energy Policy. 1999;**27**(11):679-689. DOI: 10.1016/S0301-4215(99)00025-7

[22] Vivoda V. Natural gas in Asia: Trade, markets and regional institutions. Energy Policy. 2014a;**74**:80-90. DOI: 10.1016/j.enpol.2014.08.004

[23] Cabalu H. Indicators of security of natural gas supply in Asia. Energy Policy. 2010;**38**(1):218-225. DOI: 10.1016/j.enpol.2009.09.008

[24] Hartley RP. Long-term LNG contracts. IAEE 36th International Conference Proceedings. 2013

[25] Gkonis KG, Psaraftis HN. The LNG market and a game theory approach to competition in LNG shipping. Maritime Economics & Logistics. 2009;**11**(2):227-246. DOI: 10.1057/mel.2009.2

[26] Vivoda V. LNG import diversification in Asia. Energy Strategy Reviews. 2014b;**2**:289-297. DOI: 10.1016/j.esr.2013.11.002

[27] Turner C, Johnson D. Global Infrastructure Networks. Elgar; 2017

[28] Hashimoto S. A new idea for LNG trade: Enhancing market competition through a tanker-based trading system. Journal of Asian Energy Studies. 2020;**4**:9-19. DOI: 10.24112/jaes.040002

[29] Wang N, and Mogi G. Deregulation and utility innovation: The case of Japanese electric sector. The 40th IAEE International Conference Proceedings. 2017

[30] Hashimoto S. Why Doesn't Japan have a natural gas pipeline network? Consideration from the determinant of the choice between LNG tank trucks and pipelines. International Journal of Energy Economics and Policy. 2021;**11**(3):346-353. DOI: 10.32479/ijeep.11049

[31] Baumol WJ, Oates WE. The Theory of Environment Policy. Prentice-Hall, Inc; 1975

[32] Sharkey WW. The Theory of Natural Monopoly. Cambridge University Press; 1982